ARCO

PHYSICS MCAT

STUDYGUIDE

~Allan Hilgendorf~

Stefan Bosworth, Ph.D. Marion A. Brisk, Ph. D.
Ronald P. Drucker, Ph.D. Denise Garland, Ph.D.
Edgar M. Schnebel, Ph.D. Rosie M. Soy, M.A.

PRENTICE HALL Englewood Cliffs, NJ 07632

Acquisitions Editor: *Wendy Rivers*
Project Manager: *Amy Jolin*
Manufacturing Coordinator: *Trudy Pisciotti*

SUPERCOURSE FOR THE MCAT, Second Edition by Stefan Bosworth, Marion A. Brisk, Ronald P. Drucker, Edgar M. Schnebel, Denise Garland, Rosie Soy

Printed in the United States of America

10 9 8 7 6 5 4 3 2 1

ISBN 0-13-190398-5

PRENTICE-HALL INTERNATIONAL (UK) LIMITED, LONDON
PRENTICE-HALL OF AUSTRALIA PTY. LIMITED, SYDNEY
PRENTICE-HALL CANADA INC. TORONTO
PRENTICE-HALL HISPANOAMERICANA, S.A., MEXICO
PRENTICE-HALL OF INDIA PRIVATE LIMITED, NEW DELHI
PRENTICE-HALL OF JAPAN, INC., TOKYO
SIMON & SCHUSTER ASIA PTE. LTD., SINGAPORE
EDITORA PRENTICE-HALL DO BRASIL, LTDA., RIO DE JANEIRO

Note to the Student

This study guide for physics is an excerpt from <u>MCAT SuperCourse</u> produced by ARCO and Prentice Hall Consumer Group, copyright 1991 by Stefan Bosworth, Ronald Drucker, Edgar Schnebel, Denise Garland, Rosie Soy and Marion A. Brisk. All of the physics related material has been condensed into this more manageable format, and is cross-referenced to <u>Physics</u> by Douglas C. Giancoli. Marginal notes reference each topic in this study guide to the appropriate chapter and section of the Giancoli text.

Study Well and Good Luck!

The Editorial Staff of Prentice Hall

Physics

4.1 Units

Physical quantity is a value used to numerically describe or define a physical phenomenon. This is done by first establishing a standard value and assigning it a unit or dimension. For example, the physical quantity of **length** can be defined (i.e. measured) in terms of the standard unit of the meter. A **numerical quantity** can be completely described by a pure number without any units: for example, 63.5. A **dimensional quantity** requires a numerical component and a unit component: for example, 63.5 inches. A measured dimensional quantity has different numerical values in different unit systems. Thus, 63.5 inches equals 1.61 meters. **Units** are the standard definitions against which physical quantities are measured.

Sec. 1-5
page 8

A. Common Unit Systems

There are three common unit systems encountered in physics problems—the SI, cgs and British systems.

The **SI system** is the abbreviation for Système International d'Unités. It is a metric system, also referred to as the **mks system** because the basic units of length, **mass,** and **time** are the meter, kilogram and second respectively. This is the most common system used in physics.

The **cgs system** is an older metric system that was replaced by the SI system. It uses the centimeter, gram and second as its basic units of length, mass and time.

The **British system,** also called the **English** or the **Engineering system,** is a nonmetric system. Its basic units of length, mass and time are the foot, slug and second respectively.

The conversions for length and for mass among the three systems are listed below.

Length and Mass Conversions

SI	cgs	British
1 m =	100 cm =	3.28 ft
10^{-2} m =	1 cm =	3.28×10^{-2} ft
0.305 m =	30.5 cm =	1 ft
1 kg =	10^3 g =	6.85×10^{-2} slug
10^{-3} kg =	1 g =	6.85×10^{-5} slug
14.6 kg =	1.46×10^4 g =	1 slug

1

B. Basic Units

There are seven basic physical quantities, each with an assigned unit that defines the property. The values of these units are established by international treaty. All other physical quantities are measured using units that are derived from some combination of the basic seven.

The seven basic physical quantities with their symbols and the corresponding SI units with their symbols are as follows.

See Table 1-2
page 10 for
similar list.

Physical Property	SI Unit
length, d	meter, m
mass, m	kilogram, kg
time, t	second, s
electrical charge, q	coulomb, C
temperature, T	degree kelvin, K
amount of material, n	mole, mol
luminous intensity, I	candela, cd

C. Derived Units

All other units are defined in terms of the basic units. For example, velocity is length per unit time.

$$\text{velocity} = \frac{\text{length}}{\text{time}}$$

last paragraph
page 10

Some **derived units** are given an alternative equivalent unit name and symbol. For example, the derived SI unit that measures the physical quantity force is the kilogram meter per square second, kg m/s^2, which is also called the **newton,** N.

Commonly encountered physical quantities and their derived SI units are as follows.

Each of these
derived units is
discussed when it
is first introduced
in the text, i.e.
acceleration units
are defined:
page 22.

Physical Quantity	Derived Unit	Equivalent Unit
acceleration, a	m/s^2	
area, A	m^2	
capacitance, c	C^{2}s^2/kg m^2	farad, f
electric current, I	C/s	ampere, a
electric resistance, R	kg m^2/C^2 s	ohm, Ω
electromotive force, emf	kg m^2/C s^2	volt, V
energy, E	kg m^2/s^2	joule, J
force, F	kg m/s^2	newton, N
frequency, ν	s^{-1}	hertz, hz
potential difference, V	kg m^2/C s^2	volt, V
power, P	kg m^2/s^3	watt, W
velocity, v	m/s	
volume, V	m^3	
work, W	kg m^2/s^2	newton meter

Notice that the derived units for energy and work are the same. Also identical are the derived units for emf and potential difference.

Sec. 2-2
"Changing
Units"
page 17 best
describes this
process.

D. *Dimensional Analysis*

Dimensional analysis is a procedure used to convert one unit into an equivalent unit.

Step 1. Start with a definition. A definition is a mathematical equation that relates two equivalent quantities such as:

$$A = B$$

EXAMPLE:

1 inch = 2.54×10^{-2} m

The definition can relate equivalent units of the same unit system or equivalent units in different unit systems.

Appendix B
"Dimensional
Analysis"
discusses a
different process
involved in
checking the
correctness of an
equation.

Step 2. Rewrite each definition into two **conversion factors,** CF. A CF is the ratio of the two terms, A and B, in the definition:

$$\frac{A}{B} \quad \text{and} \quad \frac{B}{A}$$

such as:

$$\frac{1 \text{ in}}{2.54 \times 10^{-2} \text{ m}} \quad \text{and} \quad \frac{2.54 \times 10^{-2} \text{ m}}{1 \text{ in}}$$

Step 3. Use the appropriate CF to convert from one unit into the other:

$$\text{Convert from } A \text{ to } B: \frac{(A)B}{A} = \frac{AB}{A} = B$$

The like terms A in the numerator and denominator cancel to leave only the B term. Similarly:

$$\text{Convert from } B \text{ to } A: \frac{(B)A}{B} = \frac{BA}{B} = A$$

such as:

$$(63.5 \text{ in}) \frac{2.54 \times 10^{-2} \text{ m}}{1 \text{ in}} = 1.61 \text{ m}$$

and

$$(1.61 \text{ m}) \frac{1 \text{ in}}{2.54 \times 10^{-2} \text{ m}} = 63.5 \text{ in}$$

Part of the strength of dimensional analysis is that you don't need to intimately understand the units of the second system as long as you have a definition that relates them to the unit system you do understand. Dimensional analysis has two component parts:

1. The numerical component is the pure number part of the quantity. Such components are arithmetically manipulated.

2. The unit component is the nonnumerical part of the term (inches, meters, etc). Such components are manipulated algebraically.

    a) Like terms multiplied together give the term raised to the appropriate power: $m \times m \times m = m^3$

    b) Like terms divided into each other cancel:

    $$\frac{m^3}{m} = m^2$$

Practice Problems

1. Kinetic energy is given by the formula $E_k = 1/2\ mv^2$, where m is the mass and v is the velocity. The derived SI unit of energy is the joule, J, which is equivalent to which combination of basic units?

   A. $kg \cdot m^2/s^2$
   B. $g\ cm^2/s^2$
   C. $g\ m^2/s^2$
   D. $kg\ m/s^2$

2. If 1 inch is equal to 2.5 cm, approximately how many inches are in 1000 kilometers?

   A. 2.5×10^4 in
   B. 4×10^4 in
   C. 2.5×10^9 in
   D. 4×10^9 in

3. A marathon race is about 26 miles long. What is the length in kilometers? (1 km = 0.623 mi.)

   A. 0.024 km
   B. 16 m
   C. 42 km
   D. 52 km

4. If a man is 6 ft 4 in tall, what is his approximate height in centimeters? (1 in = 2.54 cm)

   A. 230 cm
   B. 180 cm
   C. 150 cm
   D. 120 cm

5. The equation for the velocity of a falling drop of water is

   $$v = (2/9\ \pi)(r^2\ g\ \rho/\eta)$$

   where v is the velocity in m/s; r is the radius of the drop in m; g is the acceleration due to gravity in m/s^2; ρ is the density of the drop in kg/m^3. What must be the units of η, the coefficient of viscosity?

   A. kg m/s   C. $kg\ s^2/m$
   B. m s/kg   D. kg/m s

Techniques shown in Appendix B "Dimensional Analyses" do apply to problem 5

Answers and Explanations

1. **A** In the SI system, the units of mass, length and time are kg, m and s. This eliminates choices B and C immediately. Velocity is length over time, which is m/s in the SI system. The square of the velocity, v^2, must have the units m^2/s^2 so that $mv^2 = kg\ m^2/s^2$ which is choice A.

2. **D** This is a dimensional analysis problem:

$$10^3\ km\ (10^3\ m/1\ km)(10^2\ cm/1\ m)(1\ in/2.54\ cm)$$
$$= (1 \times 10^8/0.25 \times 10^{-1})\ in = 4 \times 10^9\ in$$

3. **C** One kilometer is less than 1 mile so the length measured in km is larger than the value measured in mi. This eliminates A and B. Choice D is eliminated because it would require 1 km to equal 1/2 mi so that each mi gave 2 km.

$$26\ mi\ (1\ m/0.623\ mi) = 42\ km$$

4. **B** 2.45 cm lies between 2 cm and 3 cm. Since 1 in = 2.54 cm you expect the height in cm to be more than twice but less than three times the corresponding value in inches:

$$76\ in\ (2\ cm/1\ in) = 152\ cm$$
and
$$76\ in\ (3\ cm/1\ in) = 228\ cm$$

The actual value is:

$$76\ in(2.54\ cm/1\ in) = 182\ cm \approx 180\ cm$$

5. **D** This is a dimensional analysis problem. Rearrange the equation so that η is alone on one side of the equal sign. Since pure numbers like 2, 9, and π have no units they can be ignored in this problem.

$$\eta = r^2\ g\ \rho/v = m^2\ (m/s^2)(kg/m^3)/(m/s)$$
$$= kg/m\ s$$

Key Words

basic units	dimensional analysis	SI system
cgs system	length	time
conversion factors	mass	
derived units	mks system	

4.2 Vectors

Sec. 2-6
page 20

A scalar quantity is a physical quantity that has only magnitude (size). It is completely described by a single number or numbers plus appropriate unit(s). **Scalar** calculations involve only ordinary arithmetic operations.

A vector quantity is a physical quantity that has both magnitude and direction. **Vector** quantities are printed in boldface (**F** = force). The magnitude of a vector is a scalar quantity (F = magnitude of force). Calculations involving vectors require vector mathematical methods.

A. Properties of Vectors

These are general properties of <u>all</u> vectors. Most of these properties are illustrated in the text in the chapter on Kinematics where 'velocity' is an important vector.

- A vector can be moved anywhere in the plane that contains it as long as the magnitude and direction of the vector are not changed in the process.

- Two vectors are identical (equal) if they have the same magnitude and direction.

- The negative of a vector is a vector with the same magnitude but opposite direction.

- A vector multiplied by a positive scalar quantity gives a vector with a different magnitude but the same direction.

- A vector multiplied by a negative scalar quantity produces a vector with a changed magnitude and an exactly reversed direction.

- Two or more vectors can be added together to give a vector sum called a resultant. The **resultant** is a single vector that can replace the other vectors acting on a body to produce the same effect as the set of vectors.

- Vector subtraction is the same as vector addition except that the negative of the vector is added.

B. Graphical Analysis

Sec. 3-1
page 43

Graphically, vector addition is accomplished by moving the vectors so that the tail of each successive vector in the addition is connected to the head of the next vector in the addition. The resultant vector **R** is the vector that connects the remaining free head to the free tail.

The order in which the vectors are added doesn't matter. All combinations must give a resultant vector of the same magnitude and direction.

If only two vectors are being added, they can be placed so that their tails are at the same origin point:

- The two vectors form the adjacent legs of a parallelogram with their resultant vector as its diagonal.

- The two vectors and their resultant also form the sides of a triangle which is equal to half the area of the parallelogram.

- If the two vectors are mutually perpendicular, the parallelogram formed is a rectangle and the two vectors and their resultant form a right triangle, as shown on the following page.

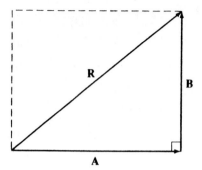

Figure 4.2-1

for part (a) see
figure 3-2
page 44

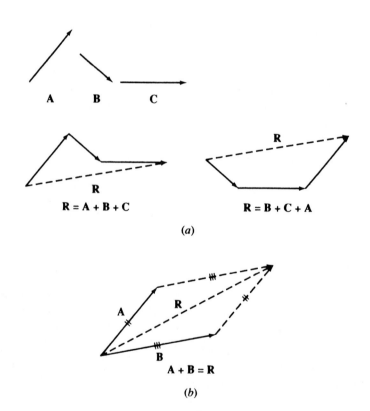

$R = A + B + C$ $R = B + C + A$

(a)

for part (b) and
(c) see figure 3-4
page 45

Pay particular
attention to the
"INCORRECT"
method shown in
the text.

$A + B = R$

(b)

Figure 4.2-2
(a) The result of the addition of a set of vectors is independent of the order in which
they are added. (b) The resultant of the addition of two vectors is the diagonal of the
parallelogram for which the two vectors are legs.

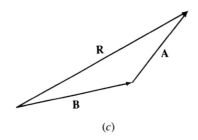

(c)

Figure 4.2-2
(c) The two vectors and their resultant from the sides of a triangle.

C. Trigonometric Analysis

The text refers to
this method
using the more
general term:
"Analytic
Method".

Every vector can be resolved into two mutually perpendicular components. The components represent the projection of the vector onto the axes of a Cartesian coordinate axis system. (Each axis of the coordinate system, x, y or z) is itself a vector. The direction of the component is specified by the axis onto which it is projected, and the component can then be given as the scalar multiplier of the unit vector for the axis. See Figure 4.2-3.

Sec. 3-3
page 46

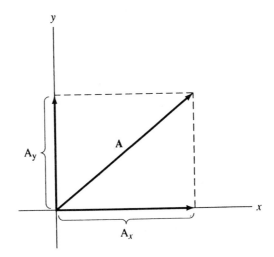

$$\mathbf{A} = \mathbf{A}_x + \mathbf{A}_y = A_x\mathbf{x} + A_y\mathbf{y}$$

Figure 4.2-3

The magnitude of a projection ranges from a minimum value of zero to a maximum value equal to that of the vector. The minimum occurs if the vector is perpendicular to the axis, the maximum occurs if the vector is collinear to an axis. Collinear vectors can either be parallel, both pointed in the same direction, or antiparallel, pointed in opposite directions. A vector and its components form the hypotenuse and legs of a right triangle, therefore, the vector can be analyzed using the trigonometric relationships of a right triangle.

Figure 3-7
page 47

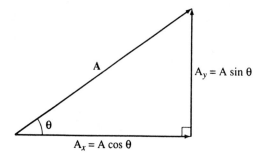

Letting theta, θ, be the angle between the vector and the x axis, the magnitude of the components of the vector can be found using the following relationships:

$$\sin \theta = \text{opposite leg/hypotenuse} = A_y/A$$

therefore, $A_y = A \sin \theta$

$$\cos \theta = \text{adjacent leg/hypotenuse} = A_x/A$$

therefore, $A_x = A \cos \theta$

Applying the Pythagorean theorem, the magnitude of the vector is found by taking the square root of the sum of the squares of the components:

$$A = (A_x^2 + A_x^2)^{1/2}$$

The angle θ can be found from any of the following relations:

Eq. 3-1
page 47

$$\sin \theta = \text{opposite leg/hypotenuse} = A_y/A$$
$$\theta = \text{arc sin } A_y/A$$

$$\cos \theta = \text{adjacent leg/hypotenuse} = A_x/A$$
$$\theta = \text{arc cos } A_x/A$$

$$\tan \theta = \text{opposite leg/adjacent leg} = A_y/A_x$$
$$\theta = \text{arc tan } A_y/A_x$$

Vector products
are not
commonly used
in a non-calculus
based physics
text.

D. *Vector Multiplication*

There are two forms of vector multiplication. One produces a scalar quantity and the other produces another vector quantity.

Dot Product

The result of this process is a scalar quantity, not a vector:

$$\mathbf{A} \cdot \mathbf{B} = AB \cos \theta$$

Notice Eq. 6-1 page 125 shows that "work" is a dot product even though the concept of a Dot Product is not introduced.

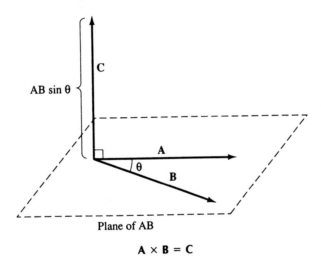

The magnitude of the projection of **B** onto **A** is given as $B_x = B \cos \theta$.

1. $\mathbf{A} \cdot \mathbf{B} = 0$ if the two vectors are perpendicular because $\theta = 90°$ and $\cos 90° = 0$.

2. $\mathbf{A} \cdot \mathbf{B} = AB$, the maximum value possible, if the two vectors are parallel because $\theta = 0$ and $\cos 0° = 1$.

Cross Product

The result of this process is a vector quantity, **C**, that is perpendicular to the plane that contained the two original vectors, **A** and **B**.

Notice Eq. 8-10c page 180 shows that "torque" is a Cross Product

$$\mathbf{A} \times \mathbf{B} = \mathbf{C}$$

The magnitude of the resultant vector **C** is always a positive number given by

$$C = AB \sin \theta$$

If **A** and **B** are parallel or antiparallel, then the angle θ between them is 0° or 180°, respectively. The sin 0° = sin 180° = 0 and the vector product is zero.

Right Hand Rule

There are always two directions that are perpendicular to a given plane. The direction of the vector product is determined by applying the right hand rule. Imagine an axis perpendicular to the plane of the two interacting vectors **A** and **B**. Let's imagine vector **B** being projected onto vector **A**. Wrap your right hand around this axis so that the fingers curl in the same direction that vector **B** would have to move in to become collinear with **A**. Your thumb will point in the direction of the resultant vector **C**.

figure 8-26
page 191

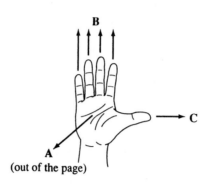

Figure 4.2-4

The "right-hand rule".

Practice Problems

1. A car travels due east for a distance of 3 miles and then due north for an additional 4 miles before stopping. What is the shortest straight line distance between the starting and ending points of this trip?

   A. 3 mi
   B. 4 mi
   C. 5 mi
   D. 7 mi

2. In the previous example, what is the angle α of the shortest path relative to due north?

   A. $\alpha = \text{arc cos } 3/5$
   B. $\alpha = \text{arc sin } 5/3$
   C. $\alpha = \text{arc sin } 4/3$
   D. $\alpha = \text{arc tan } 3/4$

3. A vector **A** makes an angle of 60° with the x axis of a Cartesian coordinate system. Which of the following statements is true of the indicated magnitudes?

   A. A_x is greater than A_y.
   B. A_y is greater than A_x.
   C. A_y is greater than A.
   D. A_x is greater than A.

4. Force is a vector quantity measured in units of newtons, N. What must be the angle between two concurrently acting forces of 5 N and 3 N respectively if the resultant vector is 8 N?

   A. 0°
   B. 45°
   C. 90°
   D. 180°

5. Two forces, **A** and **B**, act concurrently on a point C. Both vectors have the same magnitude of 10 N and act at right angles to each other. What is the closest estimate of their resultant?

   A. 0 N
   B. 14 N
   C. 20 N
   D. 100 N

Answers and Explanations

1. **C** The two legs of the trip are perpendicular, therefore the shortest distance is given by the hypotenuse of the corresponding 3-4-5 right triangle. The value can be confirmed using the Pythagorean theorem:

$$R = (3^2 + 4^2)^{1/2} = 5 \text{ mi}$$

2. **D** The angle α that gives the direction of the hypotenuse relative to the y axis is given by:

$$\tan \alpha = \text{opposite leg/adjacent leg} = 3/4$$
$$\alpha = \text{arc tan } 3/4$$

Similarly it can be given as:

$$\alpha = \text{arc cos } 4/5 \quad \text{and} \quad \alpha = \text{arc sin } 3/5$$

3. **B** Choices C and D can be eliminated immediately because the projection of a vector can never be greater than the vector itself. As the angle between the vector and the axis increases towards 90°, the magnitude of the projection decreases toward zero. The angle between the vector and the x axis is 60°, which means the angle between the vector and the y axis must be 30°. Therefore, the projection onto the y axis, A_y, must be greater than the projection onto the x axis, A_x.

4. **A** The only way the vector sum of a 3 N and a 5 N force can equal 8 N is if both forces act in the same direction. Therefore, the angle between them must be zero degrees.

$$5 \text{ N} + 3 \text{ N} = 8 \text{ N}$$

5. **B** The two vectors form the legs of a right triangle. The resultant vector is the hypotenuse. A and C are eliminated immediately because they require that the two forces be antiparallel or parallel respectively. Choice D can also be eliminated since it is too large to be the hypotenuse. The magnitude of the resultant is confirmed by using the Pythagorean theorem:

$$R = (10^2 + 10^2)^{1/2} = (200)^{1/2} = 10(2)^{1/2}$$
$$= 10(1.41) \approx 14 \text{ N}$$

Key Words

cross product
dot product
resultant

right hand rule
scalar
vector

vector product
vector component

4.3 Motion: Definition of Terms

A. Mechanics

Mechanics is the study of the relationships among matter, force and motion. It is divided into the two broad categories of statics and dynamics.

- **Statics** covers the conditions under which a body at rest remains at rest. The position of the body does not change with time.

- **Dynamics,** or kinematics, studies the laws that govern the motion of a body. The position of the body does change with time.

B. Position

Position gives the coordinate(s) for the location of a body with respect to the origin of some frame of reference. Position is a vector quantity having both magnitude and direction measured from the origin. Position is given in units of length.

Sec. 2-3
page 17

- **Frame of reference** is any co-ordinate axis system that establishes a location called the origin. Changes in the position of a body, and, therefore, in the motion of the body are measured with respect to this origin.

figure 2-2
page 17

- The **Cartesian coordinate axis system** is a three-dimensional system with three mutually perpendicular axes labelled x, y and z. This system is used to define the position or change in position of a body.

If the position of a body must be specified within a volume of space, then all three coordinate values are required to locate it. Each coordinate gives the body's position with respect to one of the three axes.

If the position or motion of a body is confined to an area of space, only two axes are required (usually x and y, by convention) to locate the body within the area. This applies to motion in a plane.

If the position of a body is further restricted to motion along a straight line, only one axis is required to locate the body. Linear motion is called translational motion. By convention, horizontal motion is along the x axis and vertical motion is along the y axis.

- Relativity of position and motion: The values for the position and motion of a body depend on the frame of reference selected. Once the frame of reference has been defined, motion is the change in position with reference to the coordinates of the chosen frame. Generally, questions concerning position, speed, velocity and acceleration use the earth as the frame of reference.

Since position is relative, motion (which is a change in position) must also be relative. Therefore, the perception of motion also depends on the frame of reference. If two bodies appear to be approaching each other, then without an external reference point it is impossible to determine if:
—both bodies are moving towards each other from opposite directions
—one body is stationary and the other is moving towards the first
—both bodies are moving in the same direction but one is moving faster and is therefore overtaking the other body.

figure 2-1
page 17

Optional Aside: If you are sitting down reading this text, how fast are you moving? This answer is not as obvious as you might expect. The earth circles the sun with an average speed of 19 meters per second, so that if your frame of reference has its origin at the sun, your speed as you sit and read is a staggering 19 m/s. If, however, the frame of reference is centered on the earth, then although both you and the planet are still revolving around the sun at 19 m/s, relative to each other neither you nor the earth is moving. You are stationary as you sit and read.

C. Motion

Motion is the change in position of a body. It is the displacement of a body with respect to bodies that are at rest (stationary) or to a fixed point (origin).

Compare figure 2-4 with figure 2-5 to show the vector nature of displacement.

- **Displacement, d,** is a vector quantity that gives the direction and magnitude of the change in the position of a body. The vector is given by the difference between the coordinates of the initial and final position.
    For linear motion (along the x axis):

$$\mathbf{d} = \Delta \mathbf{x} = x_{\text{final position}} - x_{\text{initial position}}$$

- **Distance, d,** gives the magnitude of the change in the position of a body. It is always a positive scalar quantity. The unit of distance and displacement is length.

$$d = |\Delta x|$$

- **Velocity, v,** is a vector quantity that gives the change in position as a function of time. It tells how fast the position of a body is changing in a given direction. It is the ratio of displacement to the time interval Δt during which the displacement occurs.

$$\mathbf{v} = \mathbf{d}/\Delta t$$

For linear motion this is given by:

$$\mathbf{v} = \mathbf{d}/\Delta t = \Delta \mathbf{x}/\Delta t = (x_2 - x_1)/(t_2 - t_1)$$

The units of velocity, and speed, are m/s, cm/s and ft/s in the SI, cgs and British systems respectively.
    Velocity is the slope of the curve that results from plotting the change in position vs. time.

For uniform velocity, the ratio of displacement to time is constant. The graph for uniform velocity gives a straight line. For linear motion:

$$\mathbf{v} = \mathbf{d}/\Delta t = \Delta \mathbf{x}/\Delta t = (x_2 - x_1)/(t_2 - t_1)$$

A net force must act on a body to change either its speed or its velocity. The result of this force is acceleration, **a.** The requisite for uniform linear motion is that there is no net acceleration; $\mathbf{a} = 0$.
    Nonuniform velocity: If the rate at which the body changes its position is not constant, then its velocity is nonuniform. For this to occur the body must experience a net acceleration.
    There are two broad categories for nonuniform velocity:

1. **uniform acceleration** (the velocity of the body changes at a steady rate)

2. **nonuniform acceleration** (the velocity of the body does not change uniformly)

Average velocity, v_{av}: The average velocity is the ratio of the total displacement over the total time. total displacement is the difference between the final position and the initial position and is independent of the actual path.

For a body with constant velocity (zero acceleration), the velocity of the body and its average velocity are identical:

$$\mathbf{v}_{av} = \mathbf{v} = \mathbf{d}/\Delta t = \Delta \mathbf{x}/\Delta t = \mathbf{d}/t$$

For a body with nonuniform velocity (nonzero acceleration), the total displacement vector is not necessarily the same as the vector sum for the displacement vectors of all points along the path. The average velocity over a given time interval is the numerical average of the initial and final velocity for the time interval:

$$\mathbf{v}_{av} = 1/2\ (\mathbf{v}_{final} + \mathbf{v}_{initial});\ a \neq 0$$

Sec. 2-5
page 19
Instantaneous velocity, $\mathbf{v}_{inst}$: If the velocity is nonuniform the graph of velocity *vs.* time will be a curve. The velocity at any point is the instantaneous velocity at that point. Graphically it is given by the slope of a tangent to the curve at that point.

For nonuniform velocity the value of the average velocity approaches that of the instantaneous velocity as the time interval decreases:

$$\lim_{\Delta t \to 0} \frac{\Delta \mathbf{x}}{\Delta t} = \lim_{\Delta t \to 0} \mathbf{v}_{av} = \mathbf{v}_{inst}$$

For uniform velocity the average and instantaneous velocities are identical.

- **Speed,** v, is a scalar quantity that gives the rate at which motion is occurring, which is the magnitude of the velocity of the body. It is the ratio of distance traveled to the time involved in traveling that distance.

$$v = d/\Delta t$$

Graphically, speed is the slope to the curve of distance *vs.* time.

Average speed, v_{av}, equals the total path length divided by the total elapsed time. Your speed at any point along the path may be different from this average value.

Instantaneous speed, v_{inst}, is the speed at a given moment or point along the path travelled. It is defined as the slope of the line tangent to the curve at that point.

When the speed of a body is constant (uniform), the ratio of distance to time is constant and

$$v_{inst} = v_{av} = v = d/\Delta t$$

When the rate at which a body changes position is not constant, then its speed is nonuniform and generally, $v_{inst} \neq v_{av}$.

Sec. 2-7
page 21
- **Acceleration, a,** is the vector quantity that gives the rate of change of the velocity:

$$\mathbf{a} = \Delta \mathbf{v}/\Delta t = (\mathbf{v}_f - \mathbf{v}_i)/(t_f - t_i) = (\mathbf{v}_f - \mathbf{v}_i)/t$$

When $t_i = 0$, then $\Delta t = t_f = t$.

Since $\Delta \mathbf{v} = \Delta \mathbf{x}/\Delta t$, this can also be expressed as:

$$\mathbf{a} = \Delta \mathbf{x}/\Delta t^2$$

Sec. 2-8 in the text provides a more detailed derivation of the Kinematic equations.

Therefore,

$$\mathbf{v}_f = \mathbf{v}_i + \mathbf{a}t$$

and

$$\mathbf{x}_f = \mathbf{x}_i + \frac{1}{2}\mathbf{a}t^2$$

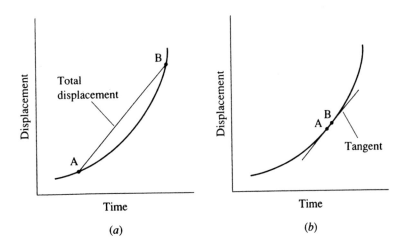

Figure 4.3-1
(a) A body with nonuniform velocity moves along the curve. The total displacement is the shortest distance from the initial point A to the final point B. (b) As the two points approach each other, the straight line connecting them becomes tangent to the curve where the two points become coincident. The average velocity and the instantaneous velocity approach the same value.

The units of acceleration are the units of velocity divided by the unit of time. In the SI system the unit is m/s². In the cgs and British systems the units are cm/s² and ft/s² respectively.

The acceleration vector can change the magnitude and/or the direction of the velocity vector. A body undergoing acceleration has a nonlinear change in position with time because the distance covered in any time interval will be changed by the acceleration.

The acceleration vector can either increase or decrease the velocity vector depending on its direction with respect to the velocity vector. It is incorrect to use the term "deceleration vector" when the acceleration acts to decrease the velocity.

Uniform acceleration: Graphically, acceleration is the slope of the curve of velocity *vs.* time. For uniform acceleration the graph of velocity *vs.* time is a linear. For nonuniform acceleration the graph is a curve and the acceleration at a given point is the tangent to the curve at that point.

Average and **instantaneous accelerations:** Like velocity, acceleration can be expressed as an average value, $\mathbf{a}_{av}$, or as an instantaneous value, $\mathbf{a}_{inst}$. For uniform acceleration, the acceleration, average acceleration and instantaneous accelerations are numerically identical since the graph of velocity *vs.* time is a straight line:

$$\mathbf{a}_{av} = \mathbf{a}_{inst} = \Delta\mathbf{v}/\Delta t = (\mathbf{v}_{final} - \mathbf{v}_{initial})/(t_{final} - t_{initial}) = \Delta\mathbf{x}/\Delta t^2$$

For nonlinear acceleration the average value approaches the instantaneous value as the time interval approaches zero:

$$\lim_{\Delta t \to 0} \frac{\Delta\mathbf{v}}{\Delta t} = \lim_{\Delta t \to 0} \mathbf{a}_{av} = \mathbf{a}_{inst}$$

Sec. 2-11
page 34

D. Graphs of Position, Velocity, and Acceleration With Respect to Time

Linear displacement is the change in the position of a body along a straight line (axis), Δx. Plots giving position, velocity and acceleration as functions of time are given on the following pages for four cases of motion.

Case 1: Body at rest; therefore velocity and acceleration are zero.

Case 2: Body moves with uniform velocity; therefore, zero acceleration.

Case 3: Body moves with nonuniform velocity and with uniform nonzero acceleration.

Case 4: Body moves with nonuniform velocity and nonuniform acceleration.

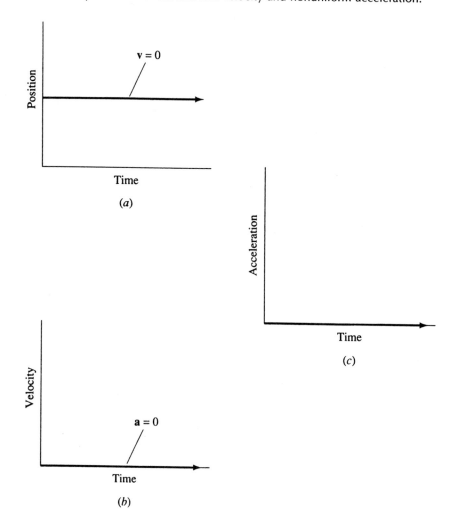

Figure 4.3-2
Case 1: For a body at rest, the graphs of (*a*) position *vs.* time, (*b*) velocity *vs.* time and (*c*) acceleration *vs.* time are all straight lines of constant zero slope.

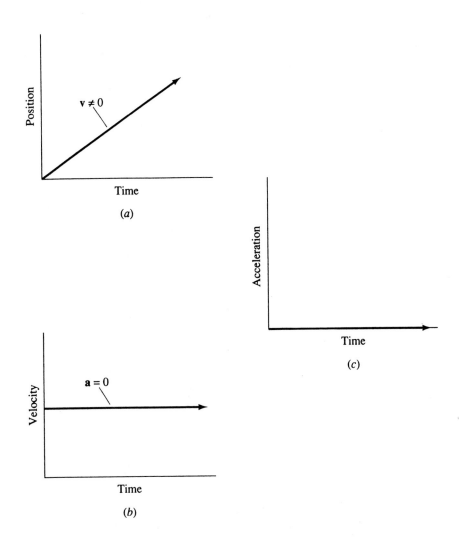

Figure 4.3-3
Case 2: For a body experiencing uniform velocity. (*a*) The graph of position *vs.* time gives a straight line with nonzero slope. The slope gives the velocity. (*b*) The graph of velocity *vs.* time gives a straight line with a zero slope. This slope is the acceleration experienced by the body. (*c*) The graph of acceleration *vs.* time is also a straight line of zero slope.

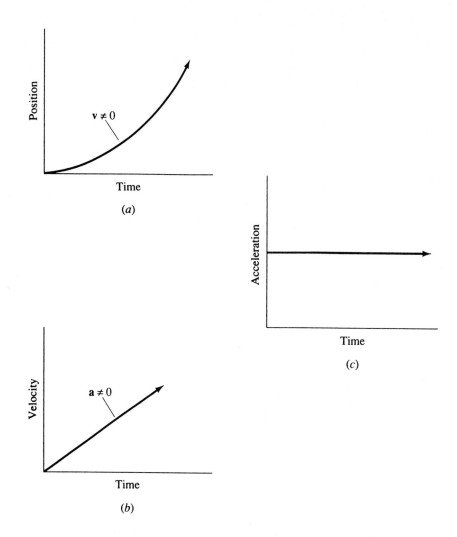

Figure 4.3-4
Case 3: A body moving with a nonuniform velocity produced by a uniform acceleration. (*a*) The graph of position *vs.* time is a curve. The instant velocity at any point is the slope of a line tangent to the curve at that point. (*b*) The graph of velocity *vs.* time is a straight line with a nonzero slope. The slope is the uniform acceleration producing the change in velocity. (*c*) The graph of acceleration *vs.* time is a straight line with a zero slope.

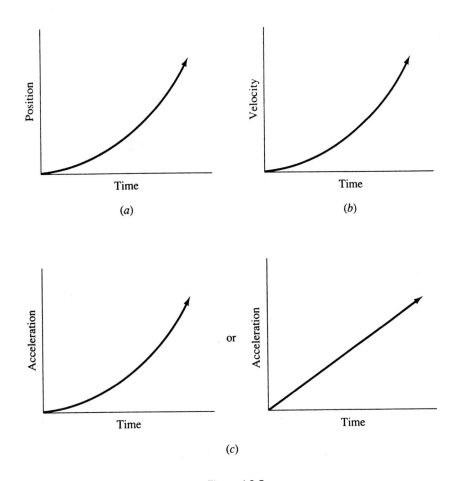

Figure 4.3-5
Case 4: The body moves with nonuniform velocity produced by a nonuniform acceleration. (a) The plot of position vs. time is a curve. The instant velocity at a given point is the slope of a tangent at that point. (b) The graph of velocity vs. time is also a curve. The instant acceleration at a given point is the slope of a tangent at that point. (c) The graph of acceleration vs. time can either be a curve or a straight line with a nonzero slope.

Practice Problems

This problem distinguishes between speed and velocity.

1. A bird flies 4.0 meters due north in 2.0 seconds and then flies 2.0 meters due west in 1.0 seconds. What is the bird's average speed?

A. 2.0 m/s
B. 4.0 m/s
C. 8.0 m/s
D. $2/3 \sqrt{5}$ m/s

2. If a car is travelling due south with a decreasing speed, then the direction of the car's acceleration is

A. due east.
B. due west.
C. due north.
D. due south.

figure 3-1
page 44

3. A hiker travels 60 meters north and then 120 meters south. What is her resultant displacement?

A. 20 m north
B. 60 m south
C. 120 m north
D. 180 m south

4. If the average velocity of a plane is 500 km per hour, how long will it take to fly 125 km?

A. 4.00 h
B. 2.00 h
C. 0.50 h
D. 0.25 h

5. Applying the brakes to a car traveling at 45 km/h provides an acceleration of 5.0 m/s^2 in the opposite direction. How long will it take the car to stop?

A. 0.40 s
B. 2.5 s
C. 5.0 s
D. 9.0 s

See figure 3-12
page 50 for an
appropriate
diagram.

6. An airplane is flying in the presence of a 100 km/h wind directed due north. What must be the velocity and heading of the plane if it is to maintain a velocity of 500 km/h due east with respect to the ground?

A. 510 km/h E
B. 300 km/h SE
C. 510 km/h SE
D. 600 km/h E

See Eq. 2-10
page 25 for any
Kinematic
problem.

7. A ball rolls down a frictionless inclined plane with a uniform acceleration of 1.0 m/s^2. If its velocity at some instant of time is 10 m/s, what will be its velocity 5.0 seconds later?

A. 5 m/s
B. 10 m/s
C. 15 m/s
D. 16 m/s

8. A ball rolls down a frictionless inclined plane with a uniform acceleration of 1.0 m/s^2. If its initial velocity is 1.00 m/s, how far will it travel in 10 s?

A. 10 m
B. 12 m
C. 60 m
D. 100 m

9. Displacement is to distance as

A. position is to change in position.
B. velocity is to speed.
C. acceleration is to velocity.
D. speed is to velocity.

10. A body with an initial speed of 25 m/s accelerates uniformly for 10 seconds to a final speed of 75 m/s. What is the acceleration?

A. 3 m/s^2
B. 5 m/s^2
C. 25 m/s^2
D. 50 m/s^2

11. A body initially at rest is accelerated at 5 m/s^2 for 10 seconds. What is its final velocity?

A. 0.50 m/s
B. 2.0 m/s
C. 15 m/s
D. 50 m/s

12. Which graph represents the motion of a body accelerating uniformly from rest?

A.
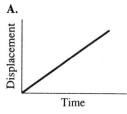

B.
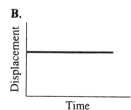

C.
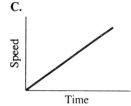

D.
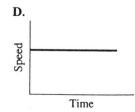

13. The speeds of a body at the ends of five successive seconds are: 180, 360, 540, 720, 900 m/h. What is the acceleration of the body?

A. 0.05 m/s^2
B. 20 m/s^2
C. 180 m/s^2
D. 180 m/h^2

Answers and Explanations

1. **A** The average speed is the ratio of the distance travelled per unit time.
$v = 6.0$ m/3.0 s = 2.0 m/s

2. **C** In order to change the magnitude but not the direction of the car's velocity, the acceleration vector must point in the opposite direction.

3. **B** Displacement is the vector that represents change in position of a body. The resultant vector is the vector sum of 60 m north and 120 m south to give 120 m south $-$ 60 m north to give 60 m south.

4. **D** The average speed is distance travelled per time required:

$t = d/v_{av} = 125$ km/500 km/h = 0.25 h

Since the plane travels 500 kilometers in one hour it should take less than one hour to fly 125 kilometers. This eliminates A and B.

5. **B** $t = (v_f - v_i)/a$. The final velocity is 0 km/h and the acceleration is -5.0 m/s^2 (The negative sign appears because the acceleration is antiparallel to the velocity). Remember to convert from kilometers to meters and from hours to seconds. Find t as follows:
$[(0 - 45$ km/h)/(-5 m/s^2)](10^3m/km)(1 h/3600 s)
$= (45 \times 10^3/5 \times 3600)$ s
$= (9 \times 10^3/36 \times 10^2)$ s
$= 2.5$ seconds

6. **C** **R** is the desired velocity, 500 km/h E, or the resultant of 510 km/h SE and the wind velocity vector 100 km/h N.

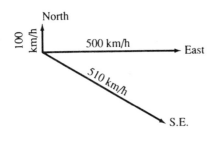

The plane's velocity must be directed SE, which eliminates choices A and D. The vector due east is the projection of the resultant and a projection cannot be greater than the vector producing it, therefore, 500 km/h is the minimum value of the vector and the choice is C.

This can be confirmed by applying the Pythagorean theorem:

$$\{(100 \text{ km/h})^2 + (500 \text{ km/h})^2\}^{1/2} = 510 \text{ km/h}$$

7. **C** $v_f = v_i + at = 10$ m/s + (1.0 m/s^2)(5.0 s)
$= 15$ m/s

8. **C** $d = v_i t + 1/2 \ at^2$
$= (1.0$ m/s)(10 s) + 1/2 (1.0 m/s^2)(10 s)^2
$= 10 + 50 = 60$ m

9. **B** Distance is the scalar equivalent of the displacement vector and speed is the scalar equivalent of the velocity vector. A and C are eliminated because all the terms are vectors. D is eliminated because the ratio is vector to scalar instead of scalar to vector.

10. **B** $a = \Delta v/\Delta t = (75$ m/s $- 25$ m/s)/10 s
$= 5$ m/s^2

11. **D** $v_f = v_i + at = at = (5$ m/s^2)10 s = 50 m/s

12. **C** Acceleration is the rate of change in velocity or speed of a body. A and B are eliminated because they are not velocity or speed *vs.* time graphs. D is eliminated because the zero slope means zero acceleration.

13. **A** $a = \Delta v/\Delta t$
$= (180$ m/h)(1 h/3600 s)/1 s
$= 0.05$ m/s^2. Remember to convert from hours to seconds.

Key Words

acceleration
average acceleration
average speed
average velocity
Cartesian coordinate
  axis system
displacement

distance
dynamics
frame of reference
instantaneous acceleration
instantaneous speed
instantaneous velocity
mechanics

nonuniform acceleration
speed
statics
uniform acceleration
velocity

4.4 Motion: Translational and in a Plane

A. Translational Motion with Uniform Acceleration

This kind of motion is the simplest type of accelerated motion. The body moves in a straight line with constant acceleration, which means the velocity of the body is changing at a uniform rate.

Uniform acceleration can be completely described by these four variables:

1. The position, **x**, or its change, $\Delta \mathbf{x} = \mathbf{x} - \mathbf{x}_0$

2. The velocity, **v**, or its change, $\Delta \mathbf{v} = \mathbf{v} - \mathbf{v}_0$

3. The time, t, or its change, $\Delta t = t - t_0$

4. The acceleration, **a**

Where $\mathbf{x}_0$, $\mathbf{v}_0$ and t_0 are the initial position, velocity and time; **x**, **v** and t are the values at some other time; **a** is the acceleration.

Questions involving uniform acceleration can be solved using the following set of six interrelated equations. The appropriate equation depends on which of the four variables are known. The pertinent missing variable(s) is listed next to the equation.

Eq. 2-10 page 25 are written in a slightly different way.

Equation	Missing Variables
1. $\mathbf{v} = \mathbf{v}_0 + \mathbf{a}t$	$\Delta \mathbf{x} = \mathbf{x} - \mathbf{x}_0$
2. $\mathbf{v}_{av} = 1/2(\mathbf{v} + \mathbf{v}_0) = \frac{1}{2}(\mathbf{v}_0 + \mathbf{a}t)$	$\Delta \mathbf{x} = \mathbf{x} - \mathbf{x}_0$
3. $\Delta \mathbf{x} = \mathbf{x} - \mathbf{x}_0 = \mathbf{v}_0 t + 1/2\, \mathbf{a}t^2$	**v**
4. $\Delta \mathbf{x} = \mathbf{x} - \mathbf{x}_0 = \mathbf{v}t - 1/2\, \mathbf{a}t^2$	$\mathbf{v}_0$
5. $\Delta \mathbf{x} = \mathbf{x} - \mathbf{x}_0 = 1/2\,(\mathbf{v} + \mathbf{v}_0)t$	**a**
6. $\mathbf{v}^2 - \mathbf{v}_0^2 = 2\mathbf{a}(\mathbf{x} - \mathbf{x}_0) = 2\mathbf{a}\Delta \mathbf{x}$	t

- **Acceleration due to gravity, g,** is a vector quantity that is always directed towards the center of the earth. The magnitude of gravity, g, for the earth in each of the three unit systems is:

SI: $g = 9.8 \text{ m/s}^2$
cgs: $g = 980 \text{ cm/s}^2$
British: $g = 32 \text{ ft/s}^2$

In a vacuum, the acceleration due to gravity is the same for all falling bodies regardless of their size or composition. This may seem counterintuitive to everyday observations until you remember that in a vacuum **g** supplies the only force acting on the body and other forces such as air resistance and buoyancy are absent. For most real (nonvacuum) conditions the effects of these other forces, while present, are assumed to be negligible.

The value of **g** is constant and doesn't change as the body falls.

Optional Aside: Actually, the effect of gravity on a body decreases as the distance between the body and the earth's center of gravity increases. However, if the distance the body falls is small compared to the radius of the earth, the variation in gravity is negligible. This is ensured by restricting free fall to points near the surface of the earth. The value of gravity also varies slightly with latitude and with the rotation of the earth, but these perturbations are negligible.

The force of gravity depends on the body producing the force, called the gravitating body. All the remarks about the acceleration of gravity due to the earth apply to the gravity due to any other body. The only difference will be the magnitude of **g.** The acceleration due to gravity near the surface of the moon is 1.67 m/s^2, while that near the surface of the sun is 274 m/s^2.

Sec 2-10
page 28

• Free-falling body: **Free fall** is an example of translational motion with uniform acceleration. The acceleration is supplied by gravity. Free fall describes the motion of bodies rising and/or falling on or near the surface of the earth.

The equations that describe free fall are identical to those for any translational motion with uniform acceleration. The vector for gravitational acceleration, **g,** replaces the acceleration vector **a.** Since vertical motion is, by convention, associated with the y axis of a coordinate system, the gravitational vector is parallel to the y axis. The vector carries a negative sign because it is always directed downward towards the negative y leg of the y axis:

$$-\mathbf{g} = \mathbf{a}_y = \text{constant}$$

Velocity in free fall calculations: The velocity is given by the equation $\mathbf{v} = \mathbf{v}_0 + \mathbf{a}t$ where the acceleration is due to gravity:

$$\mathbf{v} = \mathbf{v}_0 - \mathbf{g}t$$

The velocity at any time during free fall is the vector sum of any initial velocity, $\mathbf{v}_0$, and the change to that initial velocity due to the acceleration of gravity, **g.** This change depends on how long the acceleration has been acting on the body, t.

Position in free fall calculations: The position of a free-falling body is given by the equation $\Delta\mathbf{x} = \mathbf{x} - \mathbf{x}_0 = \mathbf{v}_0 t + 1/2\ \mathbf{a}t^2$ with the horizontal axis, x, replaced by the vertical axis, y, and the acceleration due to gravity, **g:**

$$\mathbf{y} = \mathbf{v}_0 t - 1/2\ \mathbf{g}t^2$$

figure 2-16
page 30

Free-falling body problems: The three common free-falling body problems that involve pure translational motion are outlined below.

—A body dropped from a height: This is the simplest free-falling body problem. The body starts from rest so $\mathbf{v}_0 = 0$ and the equations for velocity and position reduce to:

$$\mathbf{v} = -\mathbf{g}t \quad\text{and}\quad \mathbf{y} = -1/2\ \mathbf{g}t^2$$

Clearly, both **v** and **y** depend on the amount of time the acceleration due to gravity acts on the body.

figure 2-17
page 31

—A body thrown vertically down from a height: Here the body starts with an initial velocity directed in the same direction as the acceleration. Velocity and position are given by the equations $\mathbf{v} = \mathbf{v}_0 - \mathbf{g}t$ and $\mathbf{y} = \mathbf{v}_0 t - 1/2\mathbf{g}t^2$ respectively.

—A body thrown vertically up: This problem can be divided into three distinct regions of motion:

1. The body moves vertically upward. The velocity and position are given by the equations $\mathbf{v} = \mathbf{v}_0 - \mathbf{g}t$ and $\mathbf{y} = \mathbf{v}_0 t - 1/2\mathbf{g}t^2$. The initial velocity and the force of gravity vectors are antiparallel so that the velocity of the upward motion decreases with time and eventually becomes zero. The maximum height reached depends on the amount of time the body rises which in turn depends on the

magnitude of the initial velocity. The greater the v_0, the longer it takes for g to cancel it.

2. Eventually the effect of the gravitational acceleration cancels (overcomes) the effect of the initial upward velocity. At this point, called the zenith, the velocity becomes zero and the body stops rising.

$$v_0 = gt_{maximum}, \quad v = 0$$

Once the time ($t_{maximum}$) has been determined, the maximum height can be found:

$$y_{maximum} = v_0 t_{maximum} - 1/2 \ g \ t^2_{maximum}$$

3. At the zenith the initial velocity is zero and the only force acting on the body is the force of gravity, which starts accelerating the body towards the earth. A description of this part of the motion is identical to the case above in which a body is dropped from a height.

B. Motion in a Plane with Uniform Acceleration

A body following a curved path requires two axes to define its position at any time. By convention we use the x and y axes and the motion is said to occur in the xy plane.

Sec. 3-5
page 53
• **Projectile motion:** A body following a curvilinear path is called a projectile and the path is called its trajectory. The **range** is the horizontal distance traveled by the projectile along the x axis.

If a free-falling body has a horizontal velocity component, instead of moving only along a single vertical line it will follow the curved path of a parabola. Any vector associated with the body can be resolved into two mutually perpendicular components that are parallel to the x and y axes.

Components of the velocity vector: The initial velocity v_0 can be resolved into two components:

$$v_{0x} = v_0 \cos \theta_0 \quad \text{and} \quad v_{0y} = v_0 \sin \theta_0$$

where θ_0 is the initial angle between the velocity vector and the horizontal axis.

If the only nonnegligible force acting on the projectile is gravity, the velocity component along the x axis is constant; its value at any point will be the same as its initial value:

$$v_x = v_{0x}$$

The velocity component along the y axis feels the acceleration due to gravity. The effect of this acceleration depends on how long it has been acting on the body. Therefore, the magnitude of the component, v_y at any time is the vector sum of its initial component and the effect of the gravitational acceleration:

$$v_y = v_{0y} - gt$$

Components of the acceleration vector: The only source of acceleration is gravity, which is directed down along the y axis. The component along the x axis is zero.

$$a_y = a_{0y} = -g$$
$$a_x = a_{0x} = 0$$

Components of the Position

1. The distance or position along the x axis at any point must be the initial value plus the horizontal distance moved. The latter is the product of the horizontal velocity component and the time since the motion started.

$$\mathbf{x} = \mathbf{x_0} + \mathbf{v_{0x}}t$$

2. The height or position along the y axis has an additional term due to the acceleration of gravity:

$$\mathbf{y} = \mathbf{y_0} + \mathbf{v_{0y}}t - 1/2\ \mathbf{g}t^2$$
$$\mathbf{y} = \mathbf{y_0} + \mathbf{v_{0y}}t - 1/2\mathbf{g}t^2$$

Ex. 3-9
page 57 shows
this derivation.

Range, R, and **time of flight,** T: The range, R, is the horizontal distance between the starting and ending points of the projectile's path. The time of flight, T, is the total time the projectile spends in flight. It is twice the time required for the projectile to reach its maximum height.

$$T = 2\ \mathbf{v_{0y}}/\mathbf{g} = 2\ \mathbf{v_0}\ \sin\ \theta/\mathbf{g}$$
$$R = (\mathbf{v_0}\ \cos\ \theta)T = (\mathbf{v_0}^2/\mathbf{g})\ \sin\ 2\theta$$

Both calculations assume air resistance is negligible.

The maximum range occurs for $\theta_0 = 45°$, because $\sin 2 \times 45° = \sin 90° = 1.00$ and $R_{max} = \mathbf{v_0}^2/\mathbf{g}.$ For every range other than R_{max} there will be two angles that give the same range:

$$\theta_1 = 90 - \theta_2$$

figure 3-22 and
derivation on
page 59 shows
the path to be a
parabola.

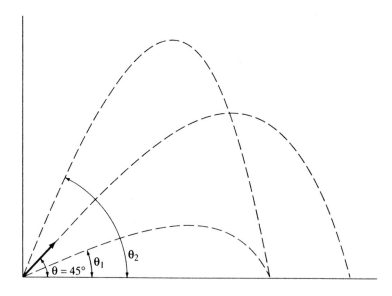

Figure 4.4-1
Diagram showing maximum range and the two equal range values produced by θ_1 and θ_2.

Summary of Position and Velocity

Equations: The initial value conditions can be modified to give values for v_x, v_y, x, and y at any point along the path by placing the origin at the starting point of the motion so that $x_0 = y_0 = 0$.

$$x = v_{0x}t = (v_0 \cos \theta_0)t$$
$$y = v_{0y}t - 1/2\ gt^2 = (v_0 \sin \theta_0)t - 1/2\ gt^2$$
$$v_x = v_0 \cos \theta_0$$
$$v_y = v_0 \sin \theta_0 - gt$$

General Trajectory Path

1. While the projectile is rising, v_y is antiparallel to gravity so its value decreases with time.

2. At the zenith, which is the maximum height reached by the projectile, $v_y = 0$.

3. On the descent, v_y and g are parallel so the vertical velocity increases.

4. The magnitude of the velocity (the instantaneous speed) of the projectile at any point is:

$$v = (v_x^2 + v_y^2)^{1/2}$$

5. The direction of the velocity at any point is:

$$\tan \theta = v_y/v_x$$

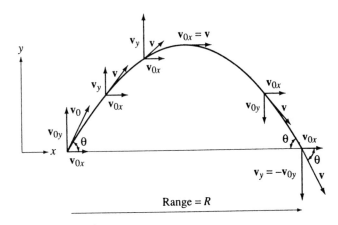

Figure 4.4-2
Diagram of the parabolic trajectory of a projectile launched with an initial velocity v_0. The velocity at any point is along the tangent at that point and can be resolved into components parallel to the x and y axes.

Sec. 5-1
page 97

• In **uniform circular motion** the speed of a body moving in a circular path is constant. However, since the direction of the body's motion is constantly changing, its velocity must be constantly changing. Furthermore, since the speed is constant, the magnitude of the velocity must be constant and it is the direction of the velocity vector that changes.

The **period,** T, is the time required to complete one revolution. If the radius of the path is R, the constant speed equals the ratio of the circumference to the period:

$$\text{constant speed} = v = 2\pi R/T$$

Figure 5-2
page 98 and
related discussion
shows the
derivation of this
equation.
In projectile motion, the acceleration is constant in both magnitude and direction. In uniform circular motion the acceleration is constant only in magnitude but not in direction. The acceleration is:

$$\mathbf{a} = \Delta\mathbf{v}/\Delta t = v^2/R = \text{speed}^2/\text{radius}$$

Circular components of acceleration: The acceleration vector can be resolved into two mutually perpendicular components with respect to the curve of the path rather than the coordinate axis system with its origin at the center of the circle.

The tangential or parallel component, $\mathbf{a}_\parallel$, at any point on the curve is coincident with a line tangent to the curve at that point. This component changes the magnitude of the velocity vector but not its direction.

The normal or perpendicular component, $\mathbf{a}_\perp$, is also called the centripetal or radial acceleration. At any point on the curve it is perpendicular to the tangent at that point and directed along a radius towards the center of the circle. This component changes the direction of the velocity vector but not its magnitude. This component must have a nonzero value if the body is to remain moving in a circular path.

A body in uniform circular motion, therefore, must have a **tangential acceleration** of zero and a normal or centripetal acceleration that is constant. This means that the body is continually "falling" (accelerating) towards the center.

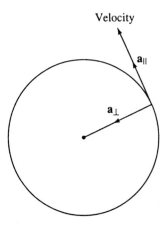

Velocity

$\mathbf{a}_\parallel$

$\mathbf{a}_\perp$

Figure 4.4-3
Components of acceleration for uniform circular motion. $\mathbf{a}_\parallel$ is collinear with the velocity vector. For uniform circular motion its magnitude is zero so it has no effect on the velocity. $\mathbf{a}_\perp$ changes the direction of the velocity. It is radially directed.

Practice Problems

1. A rock is dropped from a height of 19.6 meters above the ground. How long does it take the rock to hit the ground?

   A. 2 s
   B. 4 s
   C. 4.9 s
   D. 9.8 s

2. A spacecraft exploring a distant planet releases a probe to explore the planet's surface. The probe falls freely a distance of 40 meters during the first 4.0 seconds after its release. What is the acceleration due to gravity on this planet?

   A. 4.0 m/s^2
   B. 5.0 m/s^2
   C. 10 m/s^2
   D. 16 m/s^2

3. Which property is constant for a body in free fall?

   A. acceleration
   B. displacement
   C. velocity
   D. speed

4. Which graph represents the relationship between speed (v) and time (t) for a body falling near the surface of a planet?

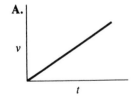
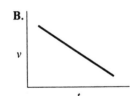

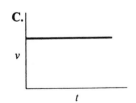
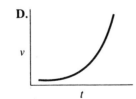

5. A ball is dropped from the roof of a very tall building. What is its velocity after falling for 5.00 seconds?

   A. 1.96 m/s
   B. 9.80 m/s
   C. 49.0 m/s
   D. 98.0 m/s

6. A quarterback throws a football with a velocity of 7 m/s and an angle of 15° with the horizontal. How far away should the designated receiver be?

   A. 1.25 m
   B. 2.50 m
   C. 5.00 m
   D. 6.25 m

Ex. 3-8
page 57

7. What must be the minimum velocity of a missile if it is to strike a target 100 meters away?

   A. 19.6 m/s
   B. 31.3 m/s
   C. 98.0 m/s
   D. 980 m/s

Ex. 3-8
page 57

8. An arrow is shot at an angle of 30° to the horizontal and with a velocity of 29.4 m/s. How long will it take for the arrow to strike the ground?

   A. 1 s
   B. 2 s
   C. 3 s
   D. 4 s

9. At what angle should a projectile be fired in order for its range to be at maximum?

   A. 30°
   B. 45°
   C. 60°
   D. 120°

10. A bomber is traveling due east with a velocity of 300 m/s when it releases a bomb. If the bomb takes 2.00 seconds to strike the ground, what was the range (horizontal distance) of the bomb's path?

   A. 100 m
   B. 150 m
   C. 300 m
   D. 600 m

Ex. 3-6
page 56

Ex. 3-6
page 56

11. A bomber is flying with a velocity of 500 m/h at an altitude of 1960 meters when it drops a bomb. How long does it take for the bomb to hit the ground?

    A. 0.50 s
    B. 20 s
    C. 50 s
    D. 200 s

Ex. 3-6
page 56

12. A ball is thrown with a horizontal velocity of 6.0 m/s. What is its velocity after 3.0 seconds of flight?

    A. 30 m/s
    B. 15.8 m/s
    C. 18 m/s
    D. 4.9 m/s

13. What is the speed of a body moving in a horizontal circle of radius 5.00 m at a rate of 2.00 revolutions per second?

    A. 7.85 m/s
    B. 15.7 m/s
    C. 31.4 m/s
    D. 62.8 m/s

Ex. 5-1
page 94

14. A 0.2-kilogram ball is tied to the end of a string and rotated in a horizontal circular path of radius 25 cm. If the speed is 10π cm/s, what is the centripetal acceleration?

    A. $2.5\ \pi$ cm/s^2
    B. $4.0\ \pi^2$ cm/s^2
    C. $0.40\ \pi$ cm/s^2
    D. $25\ \pi$ cm/s^2

15. A 70-kilogram woman standing at the equator rotates with the earth around its axis at a speed of about 500 m/s. If the radius of the earth is approximately 6×10^6 m, which is the best estimate of the centripetal acceleration experienced by the woman?

    A. 4×10^{-2} m/s^2
    B. 4 m/s^2
    C. 10^{-4} m/s^2
    D. 24 m/s^2

16. A missile is fired at an angle with the horizontal. The air resistance is negligible. The initial horizontal velocity component is twice that of the initial vertical velocity component. The trajectory of the missile is best described as

    A. semicircular.
    B. translational.
    C. parabolic.
    D. hyperbolic.

17. Which trajectory best describes a ball rolling down a curved ramp that ends at point P?

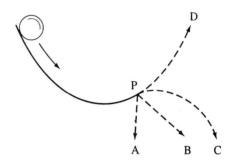

    A. A
    B. B
    C. C
    D. D

18. A cannon is placed on a flatcar of a train. The train is moving with a uniform speed 5.0 m/s when the cannon fires a ball vertically up at 10 m/s. Which vector best represents the resultant velocity of the cannon ball?

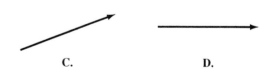

19. A bullet is fired from a gun with a muzzle velocity of 300 m/s and at an angle of 30° above the horizontal. What is the magnitude

of the vertical component of the muzzle velocity?

A. 0 m/s
B. 9.8 m/s
C. 150 m/s
D. 300 m/s

20. What is the centripetal acceleration of a 20.0 kilogram body traveling at a uniform speed of 40.0 m/s around a circular path of radius 10.0 m?

A. 320 m/s^2
B. 160 m/s^2
C. 80 m/s^2
D. 20 m/s^2

21. Car A has a mass twice that of car B. Both are traveling with the same uniform speed around a circular race track. The centripetal acceleration

A. of car A is twice that of car B.
B. of car A is half that of car B.
C. of car A is the same as that of car B.
D. of car A is four times that of car B.

22. A ball rolls with uniform speed around a frictionless flat horizontal circular track. If the velocity of the ball is doubled, the centripetal acceleration is

A. quadrupled.
B. doubled.
C. halved.
D. unchanged.

23. The centripetal acceleration of a car traveling at constant speed around a frictionless circular race track

A. is zero.
B. has constant magnitude but varying direction.
C. has constant direction but varying magnitude.
D. has varying magnitude and direction.

Answers and Explanations

1. **A** This is free fall with zero initial velocity and initial height of $y_0 = 19.6$ m. The position equation: $-y = v_0 t - 1/2\ \mathbf{g}t^2$ becomes $y = 1/2\ \mathbf{g}t^2$. Solving for time gives:

$$t = (2y_0/\mathbf{g})^{1/2} = (2 \times 19.6\ m/9.8\ m/s^2)^{1/2}$$
$$= (4\ s^2)^{1/2} = 2\ s$$

2. **B** This is a free fall problem. The object starts from rest, $v_0 = 0$. Rearrange the position equation to solve for the acceleration of gravity:

$$\mathbf{g} = 2y/t^2 = 2(40\ m)/16\ s^2 = 5.0\ m/s^2$$

3. **A** By definition a free-falling body experiences a constant acceleration due to gravity.

4. **A** A free-falling body feels a constant acceleration due to gravity. Therefore, its speed increases uniformly (linearly) as it falls. Choice B is eliminated because the speed is decreasing linearly which is opposite to the effect expected for the fall. Choice C indicates constant speed which means zero acceleration. Choice D is a curve which indicates nonlinear acceleration that causes the speed to increase in a nonlinear manner.

5. **C** The free fall velocity starting with an initial velocity of zero is:

$$v = -\mathbf{g}t = -(9.8\ m/s^2)5.00\ s = -49\ m/s$$

The velocity is 49 m/s downward (negative sign).

6. **B** Range, $R = (v_0^2/\mathbf{g})\ \sin 2\theta$
$$= \{(7\ m/s)^2/(9.8\ m/s^2)\}(0.500)$$
$$= 2.5\ m.$$

7. **B** Assume the range desired is the maximum range. It is achieved when $\theta = 45°$ so that $\sin 2\theta = \sin 90° = 1$. $R_{max} = v_0^2/\mathbf{g}$ therefore:

$$v_0 = (\mathbf{g}R_{max})^{1/2} = \{(9.8\ m/s^2)(100\ m)\}^{1/2}$$
$$= (980\ m^2/s^2)^{1/2} = 31.3\ m/s$$

8. **C** The time of flight is:

$$T = (2\ v_0 \sin \theta)/\mathbf{g}$$
$$= 2(29.4\ m/s)(0.500)/9.8\ m/s^2$$
$$= 3\ s$$

9. **B** By definition R_{max} occurs when $\theta = 45°$.

10. **D** The horizontal and vertical components of velocity are independent of each other. The initial horizontal velocity of the bomb is the same as that of the plane. The range is the horizontal distance covered which is the

product of the horizontal velocity and the time in flight:

$$x = v_0 t = (300 \text{ m/s})(2.00 \text{ s}) = 600 \text{ m}$$

11. **B** The horizontal and the vertical components of velocity are independent. The vertical component is due only to the acceleration of gravity:

$$y = 1/2 \, g t^2$$

rearranges to:

$$t = (2y/g)^{1/2} = \{2(1960 \text{ m})/(9.8 \text{ m/s}^2)\}^{1/2}$$
$$= (400 \text{ s}^2)^{1/2} = 20 \text{ s}$$

12. **A** The horizontal component of velocity is $v_x = 6.0$ m/s, and the vertical component is $v_y = gt = (9.8 \text{ m/s}^2)(3.0 \text{ s}) = 29.4$ m/s, and: $v = (v_x^2 + v_y^2)^{1/2} = \{(6.0 \text{ m/s})^2 + (29.4 \text{ m/s})^2\}^{1/2} = (900 \text{ m}^2/\text{s}^2)^{1/2} = 30$ m/s

13. **D** For a circular path, speed is $v = 2\pi R/T$. The distance covered by the body is one circumference per revolution. If the body completes 2 revolutions per second, its period (the time required to complete one revolution) must be 0.5 second.

$$v = 2\pi R/T = (2\pi)(5.0 \text{ m})/(0.5 \text{ s})$$
$$= 20 \, \pi \text{ m/s} = 62.8 \text{ m/s}$$

14. **B** The mass of the ball is not needed to find the centripetal acceleration.

$$a_c = a_\| = v^2/R = (10 \, \pi \text{ cm/s})^2/25 \text{ cm}$$
$$= 4 \, \pi^2 \text{ cm/s}^2$$

15. **A** The woman is moving in a circle whose radius is that of the earth, and moving with a constant speed equal to the rotation of the earth on its axis. The centripetal acceleration is:

$$a_c = v^2/R = (500 \text{ m/s})^2/6 \times 10^6 \text{ m}$$
$$= (25 \times 10^4 \text{ m}^2/\text{s}^2)/6 \times 10^6 \text{ m}$$
$$a_c = 4.2 \times 10^{-2} \text{ m/s}^2 \sim 4 \times 10^{-2} \text{ m/s}^2$$

16. **C** If air resistance is neglected, the trajectories of all projectiles fired near the surface of the earth follow a parabolic path because the horizontal velocity is constant while the vertical acceleration is also constant.

17. **C** The ball leaves the ramp with a horizontal component to its velocity and acquires a vertical acceleration due to gravity. The path is a typical parabolic trajectory.

18. **B** The cannon gives the ball vertical velocity while the train gives it horizontal velocity so the resultant cannot be purely vertical or purely horizontal. This eliminates choices A and D. Since the vertical component is greater than the horizontal component the resultant should make an angle greater than 45° with the horizontal. Choice B is the better description.

19. **C** The velocity has two perpendicular components, one vertically oriented and the other horizontal.

$$v_{0y} = v_0 \sin 30° = (300 \text{ m/s})(0.500)$$
$$= 150 \text{ m/s}$$

The triangle formed is a 30°-60°-90° right triangle, therefore, the side opposite the 30° angle is half the length of the hypotenuse (300 m/s).

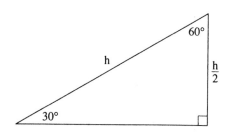

20. **B** The centripetal acceleration is the ratio of the square of the speed to the radius of the circle:

$$a_c = v^2/R = (40 \text{ m/s})^2/10 \text{ m} = 160 \text{ m/s}^2$$

21. **C** $a_c = v^2/R$. Since the speed v is the same for both cars and the radius of the track is the same, the centripetal acceleration is the same. The centripetal force on each car is different, however, because of the difference in mass.

22. **A** $a_c = v^2/R$. The acceleration and the square of the velocity are directly proportional. If you double the velocity, you quadruple the square of the velocity. Therefore, you quadruple the acceleration.

23. **B** Acceleration is always in the direction of the force producing the acceleration. In a circular path the direction of the force is continuously changing. The magnitude of the acceleration remains the same because the magnitude of the force is constant.

Key Words

acceleration due to gravity
centripetal acceleration
free fall
period

projectile motion
range
tangential acceleration
time of flight

translational motion
uniform circular motion

4.5 Force and Newton's Laws

Sec. 4-1
page 65

A. Force

Force is a vector quantity that measures the interaction between two or more bodies. A force provides an acceleration and therefore changes the motion of a body.

See the first
sentence of this
section for a good
working
definition of
force.

- Unit of force is a derived unit that defines the amount of acceleration (m/s^2), acting to move a mass (kg) some distance (m). The resulting units are summarized below:

System	Derived Unit	Basic Units
SI	newton, N	$1\ N = 1\ kg\ m^2\ s^{-2}$
cgs	dyne, dyn	$1\ dyn = 1\ g\ cm^2\ s^{-2}$
British	pound, lb	$1\ lb = 1\ slug\ ft^2\ s^{-2}$

- **Contact forces** require that the interacting bodies be in physical contact with each other. Friction is a typical contact force.

Concurrent forces are two or more forces that have lines of action that pass through some common point in the body. A set of concurrent forces can always be

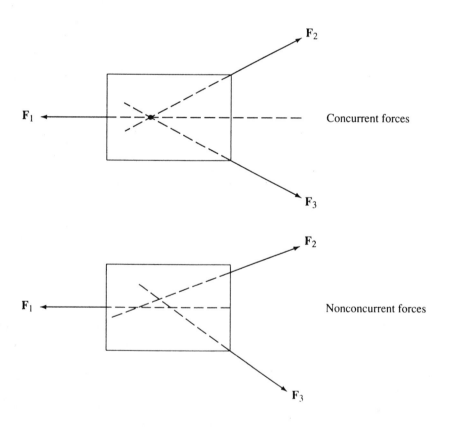

Figure 4.5-1
Concurrent and nonconcurrent forces.

resolved into a single resultant force that produces the same effect on the body as the concurrent set. The resultant tends to produce translational motion.

Nonconcurrent forces are forces that act on the same body but do not act along the same line and do not all intersect at same common point. These forces tend to cause the body to rotate.

Parallel forces are a special case of nonconcurrent forces whose lines of action never intersect. Parallel forces tend to cause the body to rotate.

- **Noncontact forces** can act through empty space without any physical contact between the interacting bodies. The force influences the motion of a body by producing a field that interacts with that body. Gravitational, electric, and magnetic forces produce fields and are typical noncontact forces.

B. Newton's First Law of Motion

Sec. 4-2
page 66

A body remains in a state of rest or of uniform motion (traveling along a straight line with constant speed) unless it is acted on by a net applied force. This first law is also called the law of inertia or the law of equilibrium.

- **Inertia** is the tendency of a body to resist a change in its motion. That is, a body will not change its motion spontaneously; some outside force must act on it.

- **Equilibrium** describes the condition where there is no net change in the motion of a system or body. A body can have forces acting on it and still be in equilibrium provided the vector sum of the forces is zero.

C. Newton's Second Law

The concept of mass is used in Newton's second law.

A net force acting on a body will cause an acceleration of the body. The acceleration vector is directly proportional to the force vector and occurs in the same direction. The proportionality constant is the **mass** of the body:

$$\mathbf{F} = m\mathbf{a}$$

See Sec. 4-3 on
page 68 for a
good definition
of mass.

1. The greater the force acting on a body, the greater the acceleration it experiences.

2. If equal forces are applied to bodies of different masses, the more massive body will experience the smaller acceleration because mass and acceleration are inversely proportional.

3. If no net force is applied to a body, the acceleration is zero and the second law gives rise to the first law.
   —A body at rest has zero velocity. Since the acceleration is also zero the body remains at rest.
   —A body in uniform motion has a constant velocity. Since the acceleration is zero the velocity remains unchanged.

Sec. 4-4
page 69

4. If a sufficient net force is applied to a body that is initially at rest, the body will start to move in the same direction as the applied force. The longer the force is applied the faster the body moves.

5. If a force is applied to a body with uniform motion, the following changes are possible:
   —If the force is in the same direction as the motion, the body will increase its speed. The angle between $\mathbf{F}$ and $\mathbf{v}$ is 0°.
   —If the force is in the opposite direction to the motion, the body will decrease its speed. The angle between $\mathbf{F}$ and $\mathbf{v}$ is 180°.

—If the force is applied at some angle other than 0° or 180° to the direction of the motion, the body will acquire nonlinear motion. That is, the direction as well as the magnitude of its motion will change.

Sec. 4-6
page 74

• **Weight, w,** is the effect of the acceleration due to gravity, **g,** on the mass of a body. Therefore weight is equal to the force of gravity on a body. It is a vector quantity that is always directed towards the center of the earth.

$$\mathbf{w} = \mathbf{F}_{grav} = m\mathbf{g}$$

Two bodies of equal mass will have different weights on different planets because the acceleration due to gravity on the two planets will be different.

Note: A common source of error in calculations occurs when the SI and cgs units for mass are incorrectly used to denote weight:

	SI	**cgs**	**British**
mass	kg	g	slug
weight	N	dyn	lb

Sec. 4-5
page 73

D. Newton's Third Law

For every action there is an equal and opposite reaction. If a body exerts a force on a second body, that second body must exert an equal counterforce on the first. The two forces form an **action-reaction pair.**

The forces of the action-reaction pair cannot act on the same body. This distinguishes them from forces that are not action-reaction pairs but happen to operate in opposite directions at some common point.

Consider a system composed of a block being pulled along a frictionless horizontal surface by a rope (Figure 4.5-2). Assume the rope is massless and the box sits on a frictionless surface. The forces acting in the system are indicated. The hand on the rope produced force $\mathbf{F}_1$ that is equal and opposite to force $\mathbf{F}_1'$ exerted by the rope on the hand. At the other end, the force $\mathbf{F}_2'$ exerted by the rope on the box is equal and opposite to the force $\mathbf{F}_2$ exerted by the box on the rope. $\mathbf{F}_1$ and $\mathbf{F}_1'$ form one action-reaction pair while $\mathbf{F}_2$ and $\mathbf{F}_2'$ form the other. Note that although $\mathbf{F}_1'$ and $\mathbf{F}_2$ are equal and opposite to each other, they do not form an action-reaction pair because they both operate on the same body, the rope.

Ex. 4-3
page 74 is a good
discussion of this
concept.

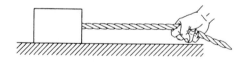

Figure 4.5-2
Forces acting on a rope and box system.

When a body rests on a horizontal surface, it exerts a force on the surface equal to the weight of the body. This force is directed downward towards the center of gravity. The surface exerts an equal upward force on the body. The force exerted by the surface is always perpendicular to the surface and is called the normal force. The weight and the normal vectors are collinear.

If the surface is inclined, the normal is still perpendicular to the surface but its magnitude is no longer equal to that of the weight of the body it supports. Its magnitude is equal to the component of the weight vector that is parallel to the normal.

Sec. 4-6
page 74

Ex. 4-4
page 76

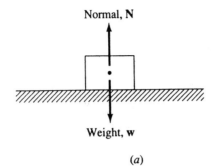

(a)

Ex. 4-14
page 88

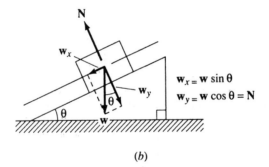

$$\mathbf{w}_x = \mathbf{w} \sin \theta$$
$$\mathbf{w}_y = \mathbf{w} \cos \theta = \mathbf{N}$$

(b)

Figure 4.5-3
Action-reaction pairs on (a) a horizontal surface and (b) an inclined plane.

Sec. 5-5
page 106

E. Newton's Law of Universal Gravitation

Every body in the universe attracts every other body with a force that is directly proportional to the product of the masses of the two bodies and inversely proportional to the square of the distance separating the two bodies. The proportionality constant is the gravitational constant, G.

$$\mathbf{F}_{gravity} = G \frac{m_1 m_2}{r^2}$$

where m_1 and m_2 are the masses of body 1 and body 2, r is the distance between them, and $G = 6.67 \times 10^{-11}$ N m^2/kg^2

The force of gravity is the gravitational attraction a body feels near the surface of a celestial body such as the earth. The gravitational forces are directed along the line that connects the centers of gravity of the two interacting bodies. For a body on or near the surface of the earth it is reasonable to assume that:

1. The height of the body above the surface is negligible compared to the radius of the earth, so $r^2 \approx r^2_{Earth}$.

2. $\mathbf{g} = Gm_{Earth}/r^2$, which means $\mathbf{g}$ depends only on the distance of the body from the center of the earth.

Sec. 4-8
page 83

F. Frictional Force

The force of friction, $\mathbf{F}_f$, is a nonconservative contact force that resists the motion of one surface sliding across another.

1. Friction acts parallel to the surfaces that are in contact.

2. Friction acts in the direction opposite to that of the net force that acts to move the body along the surface.

3. Friction depends on the identities of the materials in contact and on the conditions of their surfaces (wet, dry, oiled, rough, smooth).

   Optional Aside: The two main sources of friction are:
   • *Mechanical interactions:* Imperfections in the surfaces of the two bodies tend to interlock them together, making it more difficult for one to slide over the other.
   No surface can be made perfectly smooth, and even if it could, polishing surfaces only reduces friction to some limiting value. Polishing beyond this point actually increases the amount of friction observed.
   • *Electrical interactions:* The same interatomic and intermolecular forces that hold atoms and molecules together in solids can act between the atoms or molecules of the two sliding surfaces.

4. The frictional force is directly proportional to the force pressing the two surfaces together. This force is the normal force, $\mathbf{F}_N$. The normal force simply equals the component of the weight of the body that is perpendicular to the surface with which it is in contact unless some other force acting on the body also has a component perpendicular to the contact surfaces. As the angle of inclination of the surface increases, the value of $\mathbf{F}_N$ decreases and the frictional force decreases, making it easier for the body to slide as the incline becomes steeper.

5. The dimensionless proportionality constant for the ratio of the frictional force to the normal force is called the coefficient of friction, μ. The value of the coefficient depends not only on the identities of the two surfaces, but also on whether the body is in motion or not. For example, for two steel surfaces the static coefficient is 0.5 and the dynamic coefficient is 0.4.
   • **Static coefficient of friction,** μ_s, is the coefficient prior to the onset of motion and describes the static frictional force that helps prevent motion:

page 84
The text uses the
term "kinetic"
rather than
"Dynamic."
They mean the
same thing.

$$\mu_s = \mathbf{F}_f/\mathbf{F}_N$$

The static frictional force has a maximum value given by:

$$\mathbf{F}_{f\,max} = \mu_s\mathbf{F}_N$$

If the applied force exceeds this maximum static frictional force, motion will begin.
   • **Dynamic coefficient of friction** μ_d, is the coefficient once motion starts and describes the sliding frictional force that resists that motion. It is also called the *kinetic coefficient of friction:*

$$\mu_d = \mathbf{F}_f/\mathbf{F}_N$$

The sliding frictional force of a body is approximately independent of the body's speed or change in speed.

figure 4-25
page 85

6. For any given body on a surface, the frictional force is greater when the body is stationary than when it is in motion. This means it takes a greater force to put a body at rest into motion than to maintain the motion of a body that is already moving.

For a given pair of surfaces, the static coefficient is greater than the dynamic coefficient:

$$\mu_s > \mu_d$$

7. The frictional force is approximately independent of the size of the contact area between the two interacting surfaces. This is not a contradiction of item 5 above. The weight of the body doesn't change, so if the area supporting the weight decreases, the pressure per unit area will increase, so:

$$F_f = F_N = PA \sim constant$$

Friction on a Horizontal Plane

figure 4-24
page 84

On a horizontal surface, the normal force is exactly equal and opposite to the weight of the body, so they cancel out. The only opposition to motion is the frictional force.

To initiate motion, the net component of the applied force that is parallel to the surface, $F_{\parallel}$, must exceed the opposing static frictional force:

$$F_{\parallel} > F_f = \mu_s F_N$$

Once motion has been started, the value of the frictional force drops because the coefficient changes from its static value to its corresponding, lower, dynamic value.

To maintain uniform motion, the net component of the applied force that is parallel to the surface, $F_{\parallel}$, must equal the opposing sliding frictional force:

$$F_{\parallel} = F_f = \mu_d F_N$$

Friction on an Inclined Plane

For a horizontal surface, the normal force of a given body is constant. In an incline the normal force is not equal to the weight of the body, but only to the component of the weight force that is perpendicular to the surface of the incline. The second component is parallel to the surface of the incline and pointed downward. The frictional force opposes this component.

If the angle of inclination is changed, the value of F_N will likewise change. The minimum angle required to produce uniform speed can be determined from the fact that uniform speed means zero acceleration. Therefore, the net force in the direction of the motion must be zero. This occurs when the parallel force exactly equals the frictional force:

$$F_{\parallel}/F_N = \mu = \tan\theta$$

where θ gives the angle where the block should slide uniformly.

For a given incline, the value of the frictional force, F_f, determines the motion of the body on the incline:

Ex. 4-15
page 88

1. If the net applied parallel force is less than the frictional force, the body remains at rest on the incline.

2. If the parallel force exactly equals the frictional force, the body slides down the incline with a constant speed.

3. If the parallel force exceeds the frictional force, the body will accelerate as it slides down the incline.

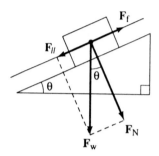

Figure 4.5-4
Forces acting on the surface of an inclined plane when friction is present. The two components are the legs of a right triangle with F_w as the hypotenuse. $F_l = F_w \sin \theta$, and $F_N = F_w \cos \theta$.

Study Sec. 4-9 "Problem Solving - A General Approach" The step of constructing a diagram for problems involving forces is most important.

Practice Problems

1. A net resultant force acting on a body will have which effect?

    A. The velocity of the body will remain constant.
    B. The velocity of the body remains constant but the direction in which the body moves will change.
    C. The velocity of the body will change.
    D. None of the above.

2. A box rests on a level table. Which of the following is an action-reaction pair of forces?

    A. The weight of the box and the upward force of the table on the box.
    B. The weight of the table and the upward force of the earth on any of the legs of the table.
    C. Both A and B.
    D. None of the above

3. Body A has a mass that is twice as great as that of body B. If a force acting on body A is half the value of a force acting on body B, which statement is true?

    A. The acceleration of A will be twice that of B.
    B. The acceleration of A will be half that of B.

    C. The acceleration of A will be equal to that of B.
    D. The acceleration of A will be one fourth that of B.

4. What is the force required to impart an acceleration of 10 m/s² to a body with a mass of 2.0 kg?

    A. 0.2 N
    B. 5 N
    C. 12 N
    D. 20 N

5. The force of gravity between two bodies is:

    A. inversely proportional to the distance between them.
    B. directly proportional to the distance between them.
    C. inversely proportional to the square of the distance between them.
    D. directly proportional to the square of the distance between them.

6. Newton's second law can be stated as which of the following?

    A. For every action there is an equal and opposite reaction.
    B. Force and the acceleration it produces are directly proportional.

C. A body at rest tends to remain at rest unless acted upon by a force.

D. None of the above

7. If two forces act concurrently on a body, the resultant force will be greatest when the angle between them is:

A. 0°
B. 45°
C. 90°
D. 180°

8. A force acting on a box can have one of the four orientations, A,B,C or D indicated below.

Review figure 3-9 page 48 for finding components of vectors.

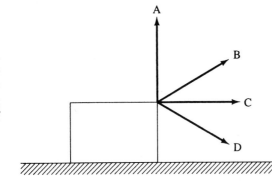

In which orientation will the force have the smallest vertical component?

A. A
B. B
C. C
D. D

9. As the vector sum of the forces acting on a body increases, the acceleration of the body must

A. increase.
B. remain the same.
C. decrease.
D. either increase or decrease depending on the direction of the resultant.

10. After a rocket is launched, its engines are shut off. The rocket continues to move in a straight line at constant speed. This is an example of

A. acceleration.
B. inertia.
C. gravitation.
D. action-reaction.

11. For a body experiencing zero acceleration, which statement is *most* true?

A. The body must be at rest.
B. The body may be at rest.
C. The body must slow down.
D. The body may speed up.

12. What is the weight of a 2.0 kg body on or near the surface of the earth?

A. 4.9 N
B. 16 lbs
C. 19.6 N
D. 64 kg m s^{-2}

13. A body accelerates at 2.5 m/s^2 when acted on by a net force of 5.0 N. The mass of the body is:

A. 0.5 kg
B. 2.0 kg
C. 12.5 kg
D. 25 kg

14. Two bodies of equal mass are separated by a distance of 2 meters. If the mass of one body is doubled, the force of gravity between the two bodies will

A. be half as great.
B. be twice as great.
C. be one fourth as great.
D. be four times as great.

15. If the mass of a moving body is doubled, the inertia of the body will be

A. half as great as its original value.
B. twice as great as its original value.
C. four times as great as its original value.
D. unchanged from its original value.

16. On a large asteroid, the force of gravity on a 10 kg body is 20 N. What is the acceleration due to gravity on the asteroid?

A. 0.5 m/s^2
B. 2.0 m/s^2
C. 9.8 m/s^2
D. 98 m/s^2

17. The distance between a spaceship and the center of the earth increases from one earth radius to three earth radii. What happens to the force of gravity acting on the spaceship?

A. It becomes 1/9 as great.
B. It becomes 9 times as great.

C. It becomes 1/3 as great.
D. It becomes 3 times as great.

18. A 100 kg astronaut lands on a planet with a radius three times that of earth and a mass nine times that of earth. The acceleration due to gravity, g, experienced by the astronaut will be

A. nine times the value of g on earth.
B. three times the value of g on earth.
C. the same value of g as on earth.
D. one third the value of g on earth.

19. For the system of a man in an elevator, his weight will appear to be greatest when

A. the elevator rises at a constant velocity.
B. the elevator is accelerated upward.
C. the elevator falls at a constant velocity.
D. None of the above because weight is constant

20. Which graph best represents the relation between the force of gravity and the mass of a free-falling body?

A.

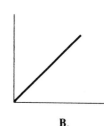

B.

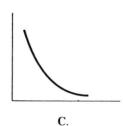

C.

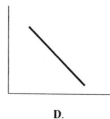

D.

Answers and Explanations

1. **C** Force produces acceleration which by definition is a change in velocity.

2. **A** The forces in an action-reaction pair must be equal and opposite and operate on two different bodies at a point of interaction. Choice C is eliminated because weight is always directed towards the center of the earth so that both forces are in the same direction. Choice B is incorrect because the force on any one leg is only equal to 1/4 the total weight of the table.

3. **D** Get the ratio of the acceleration from Newton's Second Law, $\mathbf{F} = m\mathbf{a}$.

$$\mathbf{a_A}/\mathbf{a_B} = (\mathbf{F_A}m_B/\mathbf{F_B}m_A)$$
$$= (\mathbf{F_B}/\ 2)(m_B/\ 1)(1/\ 2m_B)(1/\ \mathbf{F_B})$$
$$= 1/4$$

4. **D** Second law. $\mathbf{F} = m\mathbf{a} = (2\ \text{kg})(10\ \text{m/s}^2) = 20\ \text{N}$

5. **C** Law of universal gravitation.

6. **B** Second Law.

7. **A** Resultant of two forces is greatest when they are parallel and least when they are antiparallel.

8. **C** The force is along the horizontal. It has no vertical component.

9. **A** Second law. Force and acceleration are directly proportional.

10. **B** First law. Inertia is the tendency of a body to resist a change in its motion.

11. **B** Choices C and D are eliminated immediately because you cannot change speed or velocity without applying a force to supply the acceleration. Choice A is eliminated because the first law applies to both bodies at rest and those with uniform velocity.

12. **C** Newton's Second Law.

$$\mathbf{w} = \mathbf{F} = m\mathbf{g} = (2.0\ \text{kg})(9.8\ \text{m/s}^2) = 19.6\ \text{N}$$

13. **B** Second Law.

$$m = \mathbf{F}/\mathbf{a} = (5.0\ \text{kg m s}^{-2})/(2.5\ \text{m s}^{-2})$$
$$= 2.0\ \text{kg}$$

14. **B** Law of universal gravitation. Force is directly proportional to the product of the masses. Doubling one of the masses doubles their product and therefore doubles the force.

15. **B** Inertia.

16. **B** Second Law. $\mathbf{g} = \mathbf{F}/m$
$= (20 \text{ kg m s}^{-2})/10 \text{ kg} = 2 \text{ m/s}^2$

17. **A** Choices B and D can be eliminated because the force of gravitational attraction doesn't increase with distance.

$$\mathbf{F}_G = G \, m_1 m_2 / r^2$$

so $\mathbf{F}_G$ varies inversely with the square of the distance between the two bodies.

18. **C** $\mathbf{F}_{G \text{ earth}} = G \, m_{\text{earth}} m_{\text{body}} / r^2_{\text{earth}}$
On the planet we get:

$$\mathbf{F}_{G \text{ planet}} = G \, (9 \, m_{\text{earth}})(m_{\text{body}})/(3 r_{\text{earth}})^2 = \mathbf{F}_{G \text{ earth}}$$

Since the force of gravity and the mass of the astronaut remain the same on both the planet and on earth, the acceleration due to gravity must also be the same, $\mathbf{F}/m = \mathbf{g} = $ constant.

19. **B** Weight is mass times the acceleration due to gravity. At uniform velocity there is no acceleration so weight must remain constant. This eliminates A and C. It also eliminates D since it includes A.

20. **B** From Newton's Second Law, $\mathbf{F}/m = \mathbf{g}$. This gives the graph of a straight line with a positive slope (equal to $\mathbf{g}$).

Key Words

action-reaction pairs	friction	Newton's Third Law
concurrent forces	inertia	noncontact forces
contact forces	law of universal gravitation	parallel forces
dynamic coefficient	mass	static coefficient of friction
of friction	newton	weight
equilibrium	Newton's First Law	
force	Newton's Second Law	

4.6 Equilibrium and Momentum

A. Equilibrium

Sec. 9-1
page 205

Equilibrium describes the condition of a body experiencing no net acceleration. In order to obtain equilibrium, a body cannot experience any unbalanced or net forces.

Concurrent forces act at the same point; they supply a push or a pull to the body that can produce translational motion. Parallel forces are forces that do not act at the same point; they supply a twist or rotation to the body that can produce rotational motion.

1. In dynamic equilibrium the magnitude and direction of the velocity are constant. The body has uniform linear motion.

2. In static equilibrium the velocity is zero. A body at rest remains at rest.

3. A rigid body is one that cannot be replaced by a particle. Because an ideal particle is a point, only translational motion is possible. For a real, rigid body, rotational as well as translational motion is possible.

4. For equilibrium to occur, the body must have both translational and rotational equilibrium.

Sec. 9-2
page 206

• Translational equilibrium occurs when the resultant of all forces acting on a body is zero. If the resultant force is zero, its perpendicular components must also be zero. Newton's second law is:

$$\text{net } \mathbf{F} = \Sigma \mathbf{F} = m\mathbf{a}$$

For equilibrium, $\mathbf{a} = 0$ and we get the mathematical statement of Newton's first law: If no force acts on a body, the motion of the body will continue unchanged.

$$\Sigma \mathbf{F} = 0 \text{ which means } \Sigma \mathbf{F}_x = 0 \text{ and } \Sigma \mathbf{F}_y = 0$$

This is the condition for translational equilibrium and it is called the first condition of equilibrium.

Vector Analysis of Force

1. A force $\mathbf{F}$ can be resolved into two mutually perpendicular components, $\mathbf{F}_x$ and $\mathbf{F}_y$. The force is the resultant of the vector sum of its components.

2. Any force can be replaced by its components acting at the same point on the body.

3. If several forces act simultaneously on a body, they can be replaced by a single equivalent force called the resultant, $\mathbf{R}$, or by the components of the resultant:

$$\mathbf{R} = \Sigma \mathbf{F} \qquad \mathbf{R}_x = \Sigma \mathbf{F}_x \qquad \mathbf{R}_y = \Sigma \mathbf{F}_y$$

• *Rotation:* Parallel forces tend to cause the rotation of a body in a plane. The point around which the body rotates is called its axis of rotation. In order to determine the effects of parallel forces on the motion of a body in a given plane, a reference point called a pivot or fulcrum is selected.

Sec. 8-3
page 178

Torque, τ, is the moment of force, given by the product of the magnitude of the force producing the rotation and the torque arm:

$$\text{torque} = \text{force} \times \text{length}$$
$$\tau = \mathbf{F}r$$

The **torque arm** or lever, r, is the perpendicular distance between the line of action of the force and the axis of rotation.

The torque depends on the magnitude of the force, the point of application of the force and the direction.

The SI unit of torque is the newton-meter, N m = kg m^2/s^2.

Couples: Two parallel forces of equal magnitude that act in opposite directions in the same plane are called a couple. The torque is equal to the product of the magnitude of one of the forces and the perpendicular distance between them. A couple cannot be counterbalanced by a single force. The only way to balance a couple is with another couple.

The example problems in Sec. 9-3 page 208 are all typical statics type problems.

Rotational equilibrium occurs when the resultant torque on a body is zero.

$$\text{net } \tau = \Sigma\tau = 0 \text{ which means } \Sigma\tau_x = 0 \text{ and } \Sigma\tau_y = 0$$

The second condition of equilibrium in a plane is that the sum of the clockwise torques is equal to the sum of the counterclockwise torques about some pivot point.

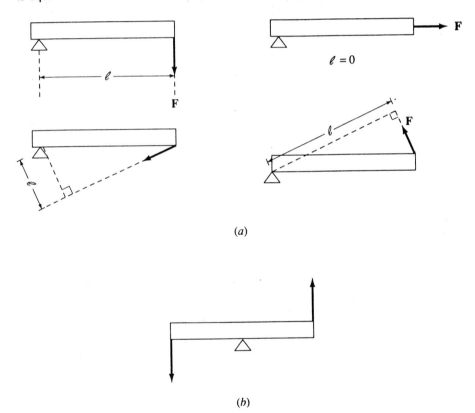

(a)

(b)

Figure 4.6-1
(a) Torques acting on a body. (b) Couple.

- **Center of mass** and **center of gravity:** A body is generally extended in space. That is, it occupies volume. However, for many calculations an extended body can be treated as if it were a particle, with no volume and all of its mass concentrated at a single point. The center of mass of a body is that point at which all of its mass can be considered to be concentrated. Similarly, the center of gravity of a body is that point at which all of its weight can be considered to be concentrated. In a uniform gravitational field the centers of mass and gravity coincide.

Optional Aside: In almost all real situations the center of gravity is slightly displaced from the center of mass, although the displacement is generally negligible. A solid cube of uniform composition rests on a table. If the mass of the cube is uniformly distributed, its center of mass is located at its geometric center. Its weight, however, is not uniformly distributed because the effect of gravity depends on the distance of the body from the center of the earth. Since the bottom half of the cube is closer to the center of the earth than the top, the bottom half weighs slightly more than the upper half. Therefore, its center of gravity lies slightly below its center of mass.

B. *Momentum*

Sec. 7-1
page 149
Momentum is the product of a force and a mass.

Momentum, **p,** is a vector quantity given by the mass of a body times its velocity. If the velocity is linear, the momentum of the body is linear. (Rotating bodies have rotational momenta.) Momentum is the force produced by a moving body:

$$\mathbf{p} = m\mathbf{v}$$

If two bodies have the same velocity, the body with the greater mass will have the greater momentum. If a small car and a big truck have the same speed, the truck will be harder to stop.

If two bodies have the same mass, the one with the greater speed will have the greater momentum. A bullet shot from a gun can do more damage to a target than a bullet thrown by hand.

Impulse is defined on page 154 in Sec. 7-3.

• **Impulse** is the product of a force and the time interval over which it acts. This is equal to the change in the momentum which the force brings about on the body:

$$\mathbf{F}\Delta t = \Delta \mathbf{p} = m\Delta \mathbf{v}$$

Sec. 7-2
page 151

• **Conservation of linear momentum:** If no outside forces act on a system of bodies, the total momentum of the system remains constant. For example, if two bodies collide, their total momentum is conserved. Therefore if one of the bodies slows down as a result of the interaction, the other must speed up:

$$m_A\mathbf{v}_{A \text{ initial}} + m_B\mathbf{v}_{B \text{ initial}} = m_A\mathbf{v}_{A \text{ final}} + m_B\mathbf{v}_{B \text{ final}}$$
$$\text{or}$$
$$m_A(\mathbf{v}_{A \text{ final}} - \mathbf{v}_{A \text{ initial}}) + m_B(\mathbf{v}_{B \text{ final}} - \mathbf{v}_{B \text{ initial}}) = 0$$

Where A and B are two bodies that undergo a change in velocity and therefore in momentum.

Newton's Third Law of Motion is a special case of the law of conservation of momentum which covers contact forces. The law of conservation of momentum also covers noncontact forces such as gravitational and electromagnetic forces.

The conservation of momentum is used to study the motions of colliding bodies.

Sec. 7-5
page 156

• In **elastic collisions,** two colliding bodies rebound without loss of kinetic energy. The sum of the kinetic energy of the bodies before the collision is equal to the sum of the kinetic energies after the collision. Perfectly elastic collisions only occur with atomic and subatomic particles.

• In inelastic collisions, two colliding bodies become joined or coupled together with a resultant loss of kinetic energy. Most real collisions are partly elastic and partly inelastic.

Kinetic energy is conserved only in perfectly elastic collisions. Momentum, however, is always conserved, whether the collision is elastic or inelastic.

Practice Problems

1. A 5.00-meter steel beam of uniform cross section and composition weighs 100 N. What is the minimum force required to lift one end of the beam?

   A. 25 N
   B. 50 N
   C. 250 N
   D. 500 N

Construct a
diagram similar
to:
Figure 9-8
page 209.

2. A nonuniform bar 8.0 meters long is placed on a pivot 2.0 meters from the lighter end of the bar. The center of gravity of the bar is located 2.0 meters from the heavier end. If a 500 N weight on the light end balances the bar, what must be the weight of the bar?

   A. 125 N
   B. 250 N
   C. 500 N
   D. 1000 N

Ex. 7-2
page 152

3. A car with a mass of 800 kg is stalled on a road. A truck with a mass of 1200 kg comes around the curve at 20 m/s and hits the car. The two vehicles remain locked together after the collision. What is their combined velocity after the impact?

   A. 3 m/s
   B. 6 m/s
   C. 12 m/s
   D. 24 m/s

4. A 1000 kg car traveling at 5.0 m/s overtakes and collides with a 3000 kg truck travelling in the same direction at 1.0 m/s. During the collision the two vehicles couple together and continue to move as one unit. What is the speed of the coupled vehicles?

   A. 2.0 m/s
   B. 4.0 m/s
   C. 5.0 m/s
   D. 6.0 m/s

Ex. 7-3
page 153

5. A rifle with a mass of 0.20 kg fires a 0.50-gram bullet with an initial velocity of 100 m/s. What is the recoil velocity of the rifle?

   A. 0.25 m/s
   B. 0.50 m/s
   C. 1.0 m/s
   D. 10 m/s

6. A 0.2-kg ball is bounced against a wall. It hits the wall with a speed of 20 m/s and rebounds elastically. What is the magnitude of the total change in momentum of the ball?

   A. 0 kg m/s
   B. 4 kg m/s
   C. 8 kg m/s
   D. 10 kg m/s

7. A tennis ball is hit with a tennis racket and the change in the momentum of the ball is 4 kg m/s. If the collision time of the ball and racket is 0.01 seconds, what is the magnitude of the force exerted by the ball on the racket?

   A. 2.5×10^{-3} N
   B. 4×10^{-2} N
   C. 3.99 N
   D. 400 N

8. A 160-pound jogger runs at a constant speed. What is his momentum if he covers 100 yards in 10 seconds?

   A. 6 ft lb /s
   B. 16 slug ft /s
   C. 100 ft lb /s
   D. 150 slug ft /s

9. How fast must a 2000-kg body travel in order to have the same momentum as a 200-g body traveling at 200 m/s?

   A. 0.02 m/s
   B. 0.05 m/s
   C. 5.0 m/s
   D. 20 m/s

10. A spacecraft with a total mass of 10,000 kg lifts off from the Kennedy Space Center in Florida. Its rockets burn for 15 seconds and produce a thrust of 3.0×10^5 N. Assuming that the net mass of the spaceship does not change as the rocket fuel is consumed, what is the final velocity of the spaceship?

    A. 2.0×10^2 m/s
    B. 4.5×10^2 m/s
    C. 3.0×10^2 m/s
    D. 5.0×10^2 m/s

11. In a combustion engine, gas is burned. The resulting explosion produces a force that drives the pistons in the engine. The force of the explosion on the piston is due to the change in momentum of the gas molecules. A 0.4-g sample of gas produces a force of 2400 N

in an explosion that lasts 10^{-3} seconds. What must be the speed of the gas molecules?

A. 6×10^3 m/s
B. 6 m/s
C. 3×10^3 m/s
D. 3 m/s

12. A 30-kg cart traveling due north at 5 m/s collides with a 50-kg cart that had been travelling due south. Both carts immediately come to rest after the collision. What must have been the speed of the southbound cart?

A. 3 m/s
B. 5 m/s
C. 6 m/s
D. 10 m/s

Construct a
diagram similar
to:
figure 9-8
page 209

13. A rod of negligible mass is 10 m in length. If a 30-kg weight is suspended from one end of the rod and a 20-kg weight from the other, where must the pivot point be placed to ensure equilibrium?

A. 4 m from the 30-kg weight
B. 4 m from the 20-kg weight
C. 6 m from the 30-kg weight
D. 5 m from the 20-kg weight

Answers and Explanations

1. **B** This is a torque arm problem. Since the beam is uniform, its center of gravity is at the midpoint of the beam, 2.50 m from either end. The weight acts at the center of gravity. Because we are lifting one end of the rod, the pivot point must be at the opposite end. The system can be pictured as:

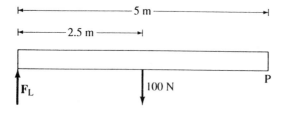

F_L is the lifting force required and it acts through a length of 5.00 m from the pivot point, P. It must counter the force of the weight acting through the length 2.50 m. At equilibrium:

$$(5.00 \text{ m})F_L = (2.50 \text{ m})(100 \text{ N})$$

The minimum force to lift the beam must be half the weight of the beam:

$$F_L = (2.50 \text{ m}/5.00 \text{ M})(100 \text{ N}) = 1/2 \ (100 \text{ N})$$

2. **B** The center of gravity occurs 4.0 m from the pivot point. At equilibrium:

$$(2.0 \text{ m})(500 \text{ N}) = (4.0 \text{ m})(\mathbf{w})$$
$$\mathbf{w} = 1/2 \ (500 \text{ N}) = 250 \text{ N}$$

3. **C** Conservation of momentum ensures that the total momentum after the collision must equal the total momentum before the collision. If A is the 800-kg car and B is the 1200-kg truck, then:

$$\text{(initial sum) } m_A v_A + m_B v_B$$
$$= m_A v_A + m_B v_B \text{ (final sum)}$$

Since the car is stalled, its initial velocity is zero and $m_A v_{A \text{ initial}}$ is zero. After collision the vehicles are joined and $v_A = v_B = v_{final}$. The equation becomes:

$$m_B v_B = (m_A + m_B) v_{final}$$

The final velocity is
$(1200 \text{ kg})(20 \text{ m/s})/(1200 \text{ kg} + 800 \text{ kg})$
$= 12$ m/s.

4. **A** Conservation of momentum gives:

$$\mathbf{p}_{\text{before collision}} = \mathbf{p}_{\text{after collision}}$$

The total momentum prior to the collision is the sum of momenta for the two vehicles. After the collision, $\mathbf{p}$ is the momentum associated with their coupled masses:

$$(1000 \text{ kg})(5.0 \text{ m/s}) + (3000 \text{ kg})(1.0 \text{ m/s})$$
$$= (4000 \text{ kg})v_{final}$$
$$v_{final} = (5000 + 3000)(\text{kg m /s})/ 4000 \text{ kg}$$
$$= 2.0 \text{ m/s}$$

5. **A** The momentum is conserved because there are no external forces acting on the rifle-bullet system during the firing. The initial momentum however is zero. The final momentum is the sum of the momentum of the bullet and the recoil momentum of the rifle:

$$\mathbf{p}_{\text{before}} = \mathbf{p}_{\text{after}}$$
$$0 = m_{\text{rifle}} v_{\text{rifle}} + m_{\text{bullet}} v_{\text{bullet}}$$
$$v_{\text{rifle}} = -(0.5 \text{ g})(100 \text{ m/s})/0.2 \times 10^3 \text{ g}$$
$$= 0.25 \text{ m/s}$$

In order to be consistent the mass of the rifle was converted into grams.

6. **C**   The change in momentum is given by:

$$\Delta p = m v_r - m v_0$$

where v_0 is the initial velocity and v_r is the rebound velocity. Since the collision is elastic the ball rebounds in the opposite direction but with the same speed it hit the wall with. Therefore, $v_r = -v_0$, and the change in momentum is:

$$\Delta p = m(-v_0) - m v_0 = -2 m v_0$$
$$= -2(0.2 \text{ kg})(20 \text{ m/s})$$
$$= -8 \text{ kg m/s}$$

The magnitude of a number is its absolute value:

$$|-8 \text{ kg m/s}| = 8 \text{ kg m/s}$$

7. **D**   Force is related to momentum by the impulse-momentum equation:

$$F t = \Delta p$$

Therefore,

$$F = \Delta p / t = (4 \text{ kg m/s})/0.01 \text{ s}$$
$$= 400 \text{ kg m/s}^2 = 400 \text{ N}$$

8. **D**   To be consistent with the British units, the distance must be converted from yards to feet and the weight in pounds into the corresponding mass in slugs.
    The units immediately eliminate A and C as possible answers.

$$m = w/g = 160 \text{ lb}/ 32 \text{ ft/s}^2 = 5 \text{ slugs}$$
$$100 \text{ yds}(3 \text{ ft/yd}) = 300 \text{ ft}$$
$$p = mv = (5 \text{ slugs})(300 \text{ ft}/ 10 \text{ s})$$
$$= 150 \text{ slug ft/s}$$

9. **A** $p_{\text{body 1}} = p_{\text{body 2}}$, therefore,

$$v_1 = m_2 v_2 / m_1$$
$$= (200 \times 10^{-3} \text{ kg})(200 \text{ m/s})/2000 \text{ kg}$$
$$= 0.02 \text{ m/s}$$

In order to be consistent, the masses of the two bodies must be in the same units, either both in grams or both in kilograms.

10. **B**   From the impulse-momentum equation, $F t = m v$ or $F \Delta t = m \Delta v$:

$$F \Delta t / m = \Delta v = v_{\text{final}} - v_{\text{initial}} = v_{\text{final}}$$

Since the spacecraft started from rest, $v_{\text{initial}} = 0$. Substitution gives:

$$v_{\text{final}} = (3.0 \times 10^5 \text{ kg m/s}^2)(15 \text{ s})/10^4 \text{ kg}$$
$$= 450 \text{ m/s}$$

11. **C**   The impulse-momentum equation rearranges to:

$$\Delta v = F_{\text{net}} \Delta t / m$$

There are two points to consider in solving the problem.
    First is the application of dimensional analysis. Do the units agree?

$$F \Delta t / m = \text{N s/g} = (\text{kg m/s}^2)\text{s/g}$$

It is necessary to change grams into kilograms (or vice versa) in order to end up in the desired units of speed, meters/second. This eliminates choices B and D.
    Second is the value of the change in velocity. Since the collisions are perfectly elastic, the gas molecules leave with the same speed they hit with. The only difference is that their direction has been reversed. Each molecule goes from a velocity of $+v$ to one of $-v$ so that $\Delta v = +v - (-v) = 2v$. Solving the equation gives C.

12. **A**   Momentum is conserved. Since both carts come to rest the total momentum after the collision must be zero which means the total momentum prior to the collision was also zero. If the two carts are labelled A and B, then:

$$m_A v_A = m_B v_B$$
$$v_B = (30 \text{ kg})(5 \text{ m/s})/ 50 \text{ kg} = 3 \text{ m/s}$$

13. **A**   Choices B and C can be eliminated because the position of the pivot is identical for

both. D can also be eliminated because it puts the pivot at the center of the rod which would only give equilibrium if the two weights were identical. The choice must be A.

This can be confirmed as follows. For equilibrium, $\Sigma \tau = 0$ which means

$$\Sigma \tau_{\text{clockwise}} = \Sigma \tau_{\text{counterclockwise}} = 0$$

The two weights cause rotation in opposite directions, therefore, if x is the distance of the pivot from the 30-kg weight, then $10 - x$ is the distance from the pivot to the 20-kg weight, and:

$$(30 \text{ kg})(x \text{ m}) = (20 \text{ kg})(10 - x \text{ m})$$
$$50x = 200, \ x = 4 \text{ m}$$

Key Words

center of gravity

center of mass

conditions of equilibrium

conservation of linear momentum

couples

ellastic collisions

impulse

rotational equilibrium

torque

torque arm

4.7 Work and Energy

A. Work and Energy

Sec. 6-1
page 125
These are equivalent terms. **Energy** is the ability of a system or body to perform work. Work is the result of a force acting to move a body some distance, and is therefore a measure of the ability of a system or body to transfer energy usefully from one point to another.

- Energy: The two broad categories of energy are **kinetic energy,** which depends on the motion of the body, and **potential energy,** which depends on its position.
    —Energy comes in a variety of forms, such as mechanical, thermal, electrical, etc. Energy can be converted from one form into another and into work.
    —In an ideal system all of the energy transferred or converted can be used to do work.
    —In a real system only part of the total energy transferred may actually be available to do work, part being lost as thermal motion (heat) of the body.

- The law of **conservation of energy** states that energy cannot be created or destroyed. It can only be transferred from point to point or converted into some other form of energy. Energy is usually reported as a scalar quantity because we are generally interested only in the amount of energy involved.

- **Work** is the ability of a system to transfer energy usefully from one form to another or from one point to another. Work is the transference of energy, not the creation or destruction of energy.
    —In an ideal system, all of the energy transferred can be used to do work.
    —In a real system only part of the energy transferred may actually be available to do work.

Work is described as the effect of a force acting to move a body through some distance. It is given by the dot product of a force and a displacement vector. The force can be a single force or the resultant of several forces:

See the explanation of Eq. 6-1 on page 125 of the text.

$$W = \mathbf{F} \cdot \mathbf{d} = Fd \cos \theta$$

where F and d are the magnitudes of the two vectors and θ is the angle between them.

The physical significance of the dot product is that the amount of work a force can produce depends on the magnitude of its component in the direction of the displacement.

The maximum amount of positive work—that is, work done by the system— occurs when the force is parallel to the displacement. Then $\theta = 0°$ and $\cos 0° = +1$ so that $W = Fd = F\Delta d$ where Δd is the change in displacement.

The maximum amount of negative work—that is, work done on the system— occurs when the force is antiparallel to the displacement. Then $\theta = 180°$ and $\cos 180° = -1$ so that $W = -Fd$.

Figure 6-2
page 126
If $\mathbf{F}$ and $\mathbf{d}$ are perpendicular to each other, no work is done either by or on the system because $\theta = 90°$ and $\cos 90° = 0$, so that $W = 0$. This observation has an interesting but for many students a counterintuitive consequence: In lifting and carrying a suitcase you supply a counter force to gravity. This force is along the vertical axis.

When you lift a suitcase, you do work on it because the force (your counter balance to gravity) is parallel to the displacement.

If you hold the suitcase, no work is done because there is no longer any displacement.

If you carry it on level ground, no work is done on or by the suitcase because F_g is perpendicular to the horizontal direction of the displacement.

Figure 15-3
page 393

Change of volume work: A common form of work is that done by an expanding or contracting gas against a constant external pressure.

$$W = -P_{ext}\Delta V = -P_{external}(V_{final} - V_{initial}) = -(F/d^2)d^3 = -Fd$$

where d is length, d^2 is area, d^3 is volume and the external pressure, P_{ext}, is the force per unit area.

- If the gas expands, the change in volume is positive so the work is negative, which means the system does work on the surroundings.

- If the gas contracts, the change in volume is negative so the work is positive, which means the surroundings do work on the system.

Work done by a
spring is also its
elastic potential
energy as shown
in figure 6-4
page 133.

Motion of a spring: When a spring is stretched or compressed and then released, a restoring force tends to return it to its equilibrium (rest) position after it oscillates about this equilibrium position. Work can be done by an oscillating spring:

$$W = -1/2\ kx^2$$

where k is the spring constant which measures how difficult it is to compress or stretch a particular spring, and x gives the displacement from the equilibrium position.

- Units of energy and work are identical. However, for convenience or for historical reasons, units for the two quantities are sometimes reported in different but equivalent units.

  In the SI system the derived unit of energy is the **joule,** J. The equivalent unit of work is the newton-meter, N m. These are related to the basic units by:

$$1\ J = 1\ N\ m = 1\ kg\ m^2\ s^{-2}$$

- In the cgs system the derived unit of energy is the erg. The equivalent unit of work is the dyne-centimeter, dyn cm. These are related to the basic unit by:

$$1\ erg = 1\ dyn\ cm = 1\ g\ cm^2\ s^{-2}$$

- In the British system the unit of energy and work is the foot-pound, ft-lb.

- Two other commonly encountered energy/work units are the calorie, cal, frequently used to measure thermal energy; and the electron volt, eV, used to measure electronic energy.

- The energy units have the following conversion relationships:

$$1\ J = 10^7\ erg = 0.738\ ft\ lb$$
$$1\ cal = 4.184\ J$$
$$1\ eV = 1.60 \times 10^{-19}\ J$$

Sec. 6-3
page 128

B. Kinetic Energy

Kinetic energy, E_k, is the energy associated with the motion of a body or the particles (atoms and molecules) composing the body. For translational motion kinetic energy is:

$$E_k = 1/2\ mv^2$$

where m is the mass and v is the velocity (vector) or speed (scalar) of the body or particle.

Macroscopic systems are those that are large enough that they obey classical or Newtonian physics. At the macroscopic level, E_k is the translational motion of the body as a whole.

Microscopic systems are those that are too small to follow classical laws and are described in terms of quantum mechanics. At the microscopic level, E_k is the change in the motion of the particles and is observed as a change in the thermal motion or temperature of the body.

Changes in kinetic energy can change the thermal motion of the particles without changing the translational motion of the body as a whole. Thus, changes in the E_k of a system are not always obvious at a macroscopic level.

also called the Work-Energy Principle

- **Work-energy Theorem:** Work can be performed by changing a system's potential energy, kinetic energy, or both. The work-energy theorem applies only to systems where the energy changes are restricted to kinetic energy. The amount of work done by the system is equal to the change in its kinetic energy.

The last part of this equation should be deleted.

$$W = \Delta E_k = E_{k\ final\ state} - E_{k\ initial\ state}$$
$$= 1/2\ mv^2_{final\ state} - 1/2\ mv^2_{initial\ state} = 1/2\ m\Delta v^2$$

Look carefully at Sec. 6-3 page 128 to see the derivation of Eq. 6-2a on page 124.

If $E_{k\ final\ state} > E_{k\ initial\ state}$, the system has gained energy and ΔE_k is positive. Therefore, work is positive. This means work is done on the system and **F** and **d** have parallel components.

If $E_{k\ final\ state} < E_{k\ initial\ state}$, the system has lost energy and ΔE_k is negative. Therefore, work is negative. This means work is done by the system and **F** and **d** have antiparallel components.

Optional Aside: Interesting consequences of the work-energy theorem include:

See Ex. 6-4 on page 130 to see how the Work-Energy principle is applied.

Free fall: A body drops under acceleration of gravity; that is, the force comes from mg, the weight of the body. The force acts in the same direction as the motion (downward). Therefore, work is positive and E_k increases as the body falls.

Vertical projectile: For a body moving vertically upward the force acting on it is still *mg*, but it is opposite to the direction of the motion. Work is negative and E_k decreases.

Terminal velocity: As a body falls in the atmosphere it eventually reaches a terminal velocity where the force of *mg* pulling the body down equals the frictional drag of the air on the body. Since F_{grav} and F_{fric} are equal and opposite, there is no net force acting on the body once it reaches terminal velocity. No work is done and E_k remains constant.

Uniform circular motion: The speed is constant, therefore, E_k is constant and no work is done on the body by the force vector. *F* points toward the center of the circle and is therefore perpendicular to the velocity. Since force and direction of motion are 90° apart, $W = 0$.

C. Potential Energy

Sec. 6-4 page 131

Potential energy, E_p, is the energy associated with the position of a body. It is the energy "stored" or contained in a system. The most common form of potential energy is gravitational:

$$E_p = mgh$$

where *m* is the mass of the body, *g* is the acceleration due to gravity and *h* is the height of the body above the ground. **Gravitational potential energy** is the ability of the force of gravity to do work on a body.

Potential energy, like kinetic energy, is the ability of a body to do work. The work associated with a change in the gravitational potential energy of a body is called the gravitational work, W_{grav}:

Figure 6-7
page 131

$$W_{grav} = \Delta E_p = (E_{p\ final\ state} - E_{p\ initial\ state}) = mg\Delta h$$

A body can perform work by changing its position with respect to the surface of the planet. The acceleration due to gravity, g, is directed vertically down.

Ex. 6-6
page 132 shows
how only the
vertical height is
important.

A decrease in the vertical height of the body means the displacement and the acceleration are parallel. The gravitational force does work as the potential energy of the body decreases. Work is done by the system.

An increase in the vertical height of the body means the displacement and the acceleration are antiparallel. The body is working against the gravitational force. The vertical force does work on the system, and the potential energy of the body increases.

D. The Law of Conservation of Mechanical Energy

This law states that the total mechanical energy, which is the sum of the kinetic and potential energy of a system, remains constant. This means that all of the energy must remain in the system changing only in how it is distributed between the kinetic and the potential forms.

$$E_{tot} = E_k + E_p$$

In order for conservation of energy to be valid, the system must be isolated from the rest of the universe. That is, the system cannot exchange either matter or energy with the outside world. Only ideal systems are truly isolated. Real systems show deviations from conservation.

Sec. 6-5
page 133

• **Conservative forces** are forces that conform to the conservation of energy law. These are forces that have a potential energy associated with them. The change in potential energy is equal and opposite to the work done by the conservative force:

$$\Delta E_p = -W_{con}$$

Gravitational and spring forces are examples of conservative forces.

For a force to be conservative, it must have one of the following three properties. Since these properties are equivalent, compliance with any one ensures that the force complies with all three.

1. A force is conservative if it does work in a closed cycle. The work for the round trip is zero.

    If W is the work involved in going from A to B, then $-W$ must be the work involved in going from B back to A via the same path. Therefore, the total work for the round trip is zero.

2. A force is conservative if the amount of work depends only on the initial and final positions of the body and is independent of the path taken.

3. The change in kinetic energy, ΔE_k, for the round trip is zero.

figure 6-11
page 134A

• **Nonconservative forces** are forces that produce deviations from the law of conservation of energy. If a force doesn't have a potential energy associated with it, then it is a nonconservative force or dissipative force. Friction is an example of a nonconservative force.

Nonconservative forces produce nonmechanical forms of energy, such as heat which is not included in the sum. The total of mechanical and nonmechanical energy is always conserved:

$$E_{tot} = E_{mech} + E_{nonmech}$$

However, the law of conservation of energy pertains only to mechanical energy.

Sec. 6-10
page 141

E. Power

Power is a scalar that gives the rate of doing work or transferring energy into or out of a system.

If the power is constant, it is equal to the ratio of the work to the time:

$$P = W/\Delta t$$

Conversely, if the power is constant, then the product of power and time gives the amount of energy used (work done).

If the power is variable, it is given as the instantaneous power, P_{inst}, which is the dot product of the force and the velocity:

$$P = \mathbf{F} \cdot \mathbf{v} = Fv \cos \theta$$

If the force and velocity are parallel, then $P = Fv$.

The SI unit of power is J/s. Another SI unit for power is the **kilowatt,** kW. Be careful not to confuse this with kilowatt hour, kW h, which is a unit of energy.

Sec. 6-8
page 137 is a
good discussion
on problem
solving.

Problems 1, 4,
and 9 deal with
the fact that the
kinetic energy is
proportional to
the square of the
speed. This is
discussed in
paragraph #1 on
page 130.

Practice Problems

1. If the speed at which a car is traveling is tripled, by what factor does its kinetic energy increase?

   A. $3^{1/2}$
   B. 3
   C. 6
   D. 9

2. Work is measured in the same units as

   A. force.
   B. energy.
   C. momentum.
   D. power.

3. What is the magnitude of the force exerted by air on a plane if 500 kilowatts of power are needed to maintain a constant speed of 100 meters per second?

   A. 5 N
   B. 50 N
   C. 500 N
   D. 5000 N

4. What happens to the speed of a body if its kinetic energy is doubled?

   A. It is multiplied by $2^{1/2}$.
   B. It is doubled.
   C. It is halved.
   D. It is multiplied by 4.

5. What is the total amount of work done when a 100 N force pushes a box to the top of an incline that is 5.0 m long, and then a 50 N force pushes the box another 5.0 m along a horizontal surface?

   A. 5.0×10^2 J
   B. 7.5×10^2 J
   C. 1.0×10^3 J
   D. 1.5×10^3 J

6. A ball with a mass of 1.0 kg sits at the top of a 30° incline plane that is 20.0 meters long. If the potential energy of the ball is 98 J at the top of the incline, what is its potential energy once it rolls half way down the incline?

   A. 0 J
   B. 49 J
   C. 98 J
   D. 196 J

7. The rate of change of work with time is

   A.   energy.
   B.   power.
   C.   momentum.
   D.   force.

8. In the diagram, a box slides down an incline. Towards which point is the force of friction directed?

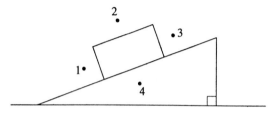

   A.   1
   B.   2
   C.   3
   D.   4

9. If the velocity of a body is doubled, its kinetic energy

   A.   decreases to half its original value.
   B.   increases to two times its original value.
   C.   decreases to one fourth its original value.
   D.   increases to four times its original value.

10. What is the velocity of a car if its engine is rated at 100 kW and provides a constant force of 5.0×10^3 N?

   A.   0.05 m/s
   B.   0.02 m/s
   C.   20 m/s
   D.   50 m/s

11. The total energy of a body free falling in a vacuum

   A.   increases.
   B.   decreases.
   C.   remains the same.
   D.   depends on the shape of the body.

12. How much work is done when a 0.50-kg mass is pushed by a 20-N force through a distance of 10.0 meters?

   A.   5 J
   B.   10 J
   C.   49 J
   D.   200 J

13. A body located 10.0 meters above the surface of the earth has a gravitational potential energy of 490 J. What is the new gravitational potential energy if the body drops to a height of 7.00 meters above the earth?

   A.   70 J
   B.   147 J
   C.   280 J
   D.   343 J

14. What is the power output by a weight lifter lifting a 10^3-N weight a vertical distance of 0.5 meters in 0.1 s?

   A.   50 W
   B.   500 W
   C.   5000 W
   D.   50000 W

15. Cart A has a mass of 1 kg and a constant velocity of 3 m/s. Cart B has a mass of 1.5 kg and a constant velocity of 2 m/s. Which statement is true?

   A.   Cart A has the greater kinetic energy.
   B.   Cart B has the greater kinetic energy.
   C.   Cart A has the greatest acceleration.
   D.   Cart B has the greater acceleration.

16. What is the kinetic energy of a 10.0-kg mass with a velocity of 2.00 m/s?

   A.   20 J
   B.   10 J
   C.   5 J
   D.   2.5 J

17. The derived unit for energy and work is called the joule, J. It is equivalent to which combination of SI units?

   A.   kg m^2
   B.   N/m^2
   C.   kg m^2/s^2
   D.   W/N

18. Two carts A and B have equal masses. Cart A travels up a frictionless incline with a uniform velocity that is twice that of Cart B. Which statement is most accurate?

    A. The power developed by A is the same as that of B.
    B. The power developed by A is half that of B.
    C. The power developed by A is twice that of B.
    D. The power developed by A is 4 times that of B.

19. The work done in raising a body must

    A. increase the kinetic energy of the body.
    B. decrease the total mechanical energy of the body.
    C. decrease the internal energy of the body.
    D. increase the gravitational potential energy.

20. A frictionless incline has a ramp length of 5.0 meters and rises to a height of 4.0 meters. How much work must be done to move a 50 N box from the bottom to the top of the incline?

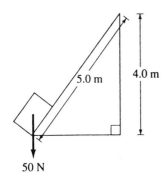

50 N

    A. 100 J
    B. 150 J
    C. 200 J
    D. 250 J

21. Work is done when a force

    I. acts vertically on a box moving along a horizontal surface.
    II. exerted on one end of a box is equal and opposite to a force exerted on the other end of the box.
    III. pushes a box up a frictionless incline.

IV. of gravitational attraction acts between a box and the surface of the earth.

    A. I and III only
    B. II and IV only
    C. I only
    D. III only

22. A 100-kilogram box is pulled 10 meters across a frictionless horizontal surface by a 50-N force. What is the change in the potential energy of the box?

    A. 0 J
    B. 2 J
    C. 20 J
    D. 50 J

23. What is the average power output of a 50-kg boy who climbs a 2.0-m step ladder in 10 seconds?

    A. 10 W
    B. 49 W
    C. 98 W
    D. 250 W

24. How much work must be done to raise a 5.0-kg block of steel from the ground to a height of 2.0 m?

    A. 2.5 N
    B. 10 N
    C. 49 N
    D. 98 N

Answers and Explanations

1. **D** The kinetic energy is directly proportional to the square of the velocity. If the velocity is tripled, then its square becomes $(3v)^2 = 9v^2$; and E_k must increase by the same nine-fold amount.

2. **B** By definition, energy is the ability to do work and shares the same units (joules).

3. **D** The average power is the product of the force acting on the plane and the speed, $P = Fv$. Solving for the required force gives:

$$F = P/v = 500 \times 10^3 \ \text{W}/(100 \ \text{m/s})$$
$$= 5 \times 10^3 \ \text{N}$$

Watch the units: W = J/s, J = kg m,
N = kg m²/s²

4. **A** E_k and v^2 are directly proportional. If E_k is doubled, then v^2 is doubled and the square root of v^2 gives $2^{1/2}$. The velocity, v, must increase to $2^{1/2}$ its original value.

5. **B** The total work done on the body is the sum of the products of the applied force and the distance moved:

$$\text{Work} = (100 \text{ N})(5 \text{ m}) + (50 \text{ N})(5 \text{ m})$$
$$= 750 \text{ N m} = 750 \text{ J}$$

6. **B** Since the height is decreasing, the E_p is decreasing and must have a value less than its initial reading of 98 J. D is eliminated because its value is higher than the initial E_p value. C is eliminated because its value is the same as the initial value of the E_p. A is eliminated because the only way to have zero E_p is if the body is at ground level. That leaves B as the only option.
   Confirm by:

$$E_p = mgh = (1.0 \text{ kg})(9.8 \text{ m/s}^2)(5.0 \text{ m}) = 49 \text{ J}$$

The 5.0 m height comes from treating the incline as a right triangle. Since sin 30° = 0.500, the length of the side opposite the 30° angle is half the length of the hypotenuse. When the ball has rolled halfway down the ramp, the hypotenuse is 10.0 m long.

7. **B** Power is the rate of doing work and is given by the ratio of the work done over the time required to do the work.

8. **C** Friction acts in the direction opposite to the motion. The force moving the box is directed towards point 1, so the resulting frictional force must be towards point 3. Point 2 is the direction of the normal; Point 4 is the direction of the weight.

9. **D** Doubling v means quadrupling v^2; therefore, E_k must increase to 4 times its original value.

10. **C** Power is force times speed, $P = Fv$, so

$$v = P/F = 100 \times 10^3 \text{ W}/5 \times 10^3 \text{ N}$$
$$= 20 \text{ m/s}$$

11. **C** Apply the law of conservation of energy. As the body falls, its E_p decreases. However, the body's E_k increases because the velocity at

which it is falling increases. Therefore, the sum of the energies remains the same.

12. **D** Since you are not given information to the contrary, assume that the force acts in the same direction as the motion. Then:

$$W = Fd = (20 \text{ N})(10.0 \text{ m/s}) = 200 \text{ J}$$

13. **D** The gravitational potential energy and the height of the body above the ground are directly proportional. Since the weight of the body (wgt = mg) remains constant, reducing the height to 7/10 of its value must reduce $E_{p \text{ grav}}$ by the same proportion. We can eliminate A and B because both are less than half the original E_p requiring that the body drop to less than half its original height. By inspection, choice C is only a little more than 1/2 and D is almost 3/4 the original value of E_p, making D the likely answer.
   Confirm this by:

$$E_{p \text{ new}} = E_{p \text{ old}} (h_{\text{new}}/h_{\text{old}}) = (490 \text{ J})(7/10)$$
$$= 343 \text{ J}$$

14. **C** Power = rate of doing work, $P = W/t = Fd/t$:

$$P = (10^3 \text{ N})((0.5 \text{ m})/0.1 \text{ s} = 5 \times 10^3 \text{ W}$$

15. **A** Since the velocity is constant the acceleration must be zero and we can eliminate choices C and D. Solving $E_k = 1/2 \, mv^2$ for the two carts gives:

$$E_{kA} = (1 \text{ kg})(9) = 9 \text{ J}$$
$$\text{and}$$
$$E_{kB} = (1.5)(4) = 6 \text{ J}$$

16. **A** $E_k = 1/2 \, mv^2 = (1/2)(10.0 \text{ kg})(2.00 \text{ m/s})^2$
    $= 20 \text{ J}$

17. **C** Work is the product of force (N = kg m/s²) and distance (m).

$$W = Fd = \text{kg m}^2/\text{s}^2$$

18. **C** Since each cart has a constant velocity, the acceleration on each must be zero. Thus, the only force acting on a cart must be the force that counters the cart's weight. Both carts have the same mass so both are subject

to the same force. Power is the product of force and velocity, so:

$$P_A/v_A = P_B/v_B = F = \text{constant}$$

Power and velocity are directly proportional. Cart A travels at twice the velocity and must develop twice the power.

19. **D** By definition, $E_{p\,grav}$ changes with the height of a body above the ground.

20. **C** Because the incline is frictionless, the work done is that associated with the change in vertical distance, 4 m. The work is that done against gravity and equals the weight of the box, $F = \text{wgt} = mg$.

$$W = Fd = (50\ N)(4\ m) = 200\ J$$

21. **D** $W = Fd$, so we can eliminate II and IV because no motion (change in distance) occurs. The force described in I is perpendicular

to the direction of the motion so I is also eliminated. Only III fits the definition of work.

22. **A** $\Delta E_p = mg\Delta h$. The height of the box didn't change so its potential energy doesn't change.

23. **C** $P = W/t$. Work is force times distance and force is mass times acceleration. The acceleration here is that of gravity. So:

$$\begin{aligned}P &= (m \times a \times d)/t\\ &= (50\ kg)(9.8\ m/s^2)(2.0\ m)/10\ s\\ &= 98\ W\end{aligned}$$

24. **D** The work done is against the force of gravity. The force of gravity is the weight of the block:

$$\begin{aligned}W &= Fd = mgh\\ &= (5.0\ kg)(9.8\ m/s^2)(2.0\ m) = 98\ N\end{aligned}$$

Key Words

conservation of energy
conservative forces
energy
gravitational potential
   energy
joule

kilowatt
kinetic energy
Law of Conservation of
   Mechanical Energy
nonconservative forces
potential energy

power
work
work-energy theorem

4.8 Solids and Fluids

A. Phases of Matter

Introduction to
chapter 10
page 237

Matter exists in three phases or states: solid, liquid and gas, distinguished by the strength of the intermolecular forces holding the component molecules together.

Note: All remarks about molecular interactions also apply to substances where the component particles are atoms instead of molecules; their interactions are due to analogous interatomic forces.

- In solids the particles are held in a fixed pattern by strong intermolecular forces, and each molecule oscillates around its fixed or equilibrium position. It is the fixed position of the constituent molecules that gives solids their rigidity. A sample of a solid has a definite volume and a definite shape which it maintains regardless of the shape of the vessel it is placed in. There are well defined boundaries or interfaces between the surface of the solid and the surrounding environment.

figure 29-21
page 788

Crystalline solids are solids in which the fixed arrangement of the molecules forms a regular and repeating pattern. This three-dimensional pattern is called the crystal lattice of the solid. Salts and minerals are examples of crystalline solids. A crystalline solid is characterized by a sharp, well defined melting point, which is the temperature at which the substance makes the transition between its solid and liquid phases.

In **amorphous solids** the positions of the molecules are fixed but the pattern of their arrangement is random. Glass is an example of an amorphous solid. An amorphous solid will not have a well defined melting point.

- Fluids are liquids and gases which are nonrigid substances where the molecules are essentially free to move about or flow.

In liquids the intermolecular forces are weaker than in solids and the molecules are farther apart. A liquid has a definite volume but not a definite shape. It assumes the shape of the vessel containing it. There is usually a single interface between the upper surface of the liquid and the environment (usually air) above it.

In a gas the molecules ideally do not interact with each other. A gas has neither a definite volume nor a definite shape but will expand to completely fill any closed vessel containing it. There is no interface between the gas sample and the surrounding environment, just the walls of the vessel.

- Molecular motion is described by the kinetic theory of matter, which states that molecules of all substances are in constant motion. The square of the velocity of the particles is directly proportional to the kinetic energy, E_k, of the substance, which in turn is directly proportional to its temperature.

In changes of phase the substance is transformed from one physical state into another. Phase changes are equilibrium phenomena, and therefore occur at a constant temperature that depends on the identity of the substance, the transition occurring, and the pressure.

Melting is the transition between the solid and liquid phases of a substance. Its reverse, from liquid to solid, is called freezing or solidification.

Boiling or vaporization is the transition from liquid to gas. The reverse transition is called condensation.

Sublimation is the change from solid to gas.

Each of these processes is reversible. If the molecular motion is great enough to overcome the attractive intermolecular forces holding the particles together, the transition will be to a less structured phase (such as solid to liquid). If the intermolecular forces overcome the molecular motion, the transition will be to a more-structured phase (such as liquid to solid).

"Cohesion and
adhesion"
page 263
Cohesion is the attractive force between molecules of the same substance. This is the force that holds like molecules together and opposes the dissolving of one substance in some other substance.

Adhesion is the attractive force between molecules of different substances. This is the force that allows one substance to become dissolved in another.

Sec. 13-13
page 362
Diffusion is the penetration of molecules of one substance into a sample consisting of a second substance. It can occur among all three phases but is most common in fluids.

Sec. 10-2
page 238
• **Pressure** is a physical quantity that gives the ratio of the magnitude of a force, F, perpendicular to a given surface area, A:

$$P = F/A$$

Pressure is an example of a physical quantity called stress.

Units of pressure: The SI system has three equivalent units of pressure, the Pascal, Pa, the Newton per square meter, N/m^2, and the kilogram per meter per square second, $kg/m\ s^2$:

$$1\ Pa = 1\ N/m^2 = 1\ kg/m\ s^2$$

Some conversions between different units of pressure are:

$$1\ atm = 760\ mm\ Hg = 760\ torr = 1.01 \times 10^5\ Pa = 14.7\ lb/in^2$$

Sec. 10-1
page 238
• **Density,** ρ, is the ratio of the mass to volume of a substance:

$$\rho = m/V$$

An SI unit of density is kg/m^3. Other units include kg/L where one liter is the volume of a perfect cube with sides of length 0.1 meters:

$$1\ L = 10^{-3}\ m^3$$

For most substances the density of the liquid is less than that of the solid so that most solids sink in liquids of the same substance. A notable exception to this observation is water, where the solid form is less dense than the liquid. This is why ice floats.

The density of water, ρ_{H_2O}, is $1\ g/cm^3$ in the cgs system and $1000\ kg/m^3$ in the SI system.

For solids and liquids the volume of a given sample at a given temperature is constant, making the density an intensive property, which means it is independent of the size of the sample. The value of the density depends on the identity of the solid or liquid substance and the temperature at which the measurements are made.

Optional Aside: The density of a solid or liquid is an intensive property of the substance that is given by the ratio of the two extensive properties: mass and volume. Intensive properties do depend on the physical size of the sample. However, the greater the mass of the sample, the greater the volume that mass tends to occupy so that the ratio remains constant.

Optional Aside: Increasing the temperature of a substance increases the thermal motion of the molecules so that they tend to move further away from each other. Therefore, the volume occupied by a solid or liquid in most cases will increase slightly with increasing temperature. Since density and volume are inversely proportional, increasing the volume occupied by a given mass will decrease its density.

Specific gravity, s.g., is also called relative density. It is the ratio of the density of a given substance to the density of water at the same temperature. Since density is the ratio of mass to volume, the specific gravity also gives the ratio of the mass of a given volume of a substance to the mass of an equal volume of water. The specific gravity is a dimensionless physical property:

$$\text{s.g.} = \rho_{body}/\rho_{H_2O}$$

Gases: Since gases expand to fill the available volume, the density of a given sample will change as you change the volume of the vessel.

For gases, density changes significantly with changes in T and P and is more correctly described as the concentration of the gas. The mass of the gas is thus uniformly distributed throughout the vessel.

B. Solids

Sec. 9-6
page 218

- **Deformation** is the change in dimension or shape of a body due to the application of an external force.

  Elasticity is the ability of a deformed body to return to its original dimensions once the external force causing the deformation is removed.

  The **elastic limit** is the minimum stress that produces an irreversible deformation of the solid. Each sample of a substance has a maximum amount of deformation that it can undergo. If it is deformed beyond this point, called the elastic limit, the sample remains deformed even after the external deforming force is removed.

  Tensile strength is the force required to actually break a rod with unit cross sectional area of the substance.

- Stress is the force acting on a solid divided by the area over which the force acts. It therefore has units of N/m^2 in the SI system. It is a physical quantity that is analogous to and has the same units as pressure. The difference is that stress gives the ratio of the internal force produced when the body is deformed to the area of the body that the force affects. There are three main types of stress. Two of them, **tension** and **compression,** are called **normal stresses** because the force is perpendicular to the area it affects.

$$\text{normal stress} = F_{\perp}/A$$

In contrast, a shear stress is a tangential stress because the force is coplanar with (parallel to) the area it affects.

Optional Aside: Deforming a substance usually changes the distance between its component molecules, which in turn changes the magnitude and/or direction of the intermolecular forces of the substance. When the external deforming force is removed, the intermolecular forces act as a restoring force, returning the molecules to their original equilibrium positions. This restoring force is the internal force used in the stress equation.

Tension or tensile stress is the effect of two external antiparallel forces with the same line of action pulling on opposite ends of a body. The two forces oppose each other and tension tends to elongate the body, increasing its length from ℓ to $\ell + \Delta\ell$ where $\Delta\ell$ is the change in length produced by the tension. Each force is normal to the area it acts on. The line of action of the forces corresponds to the axis of elongation.

Compression is the reverse of tension. It is due to the effect of two antiparallel collinear external forces pushing on opposite sides of the same body. The two forces again oppose each other but the net effect is to compress the body, decreasing its length by $\Delta\ell$ from ℓ to $\ell - \Delta\ell$. Each force is again perpendicular to the area it acts on.

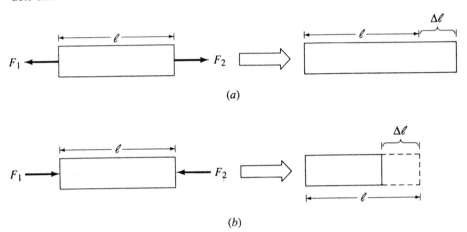

Figure 4.8-1
(a) Elongation and (b) compression of a rod under stress.

Note: For some substances the tensile and compression stresses are equal in magnitude for a given force. In other substances they may be quite different. For instance, concrete can generally withstand a larger compression force than tensile force before it breaks.

In shear stress the two antiparallel forces do not have the same line of action. The net effect is that the forces act along opposite areas instead of perpendicular to them. One surface area is displaced a horizontal distance Δx with respect to the opposite surface. Shear stress is also called tangential stress.

$$\text{shear stress} = F_{\parallel}/A$$

figure 9-23
page 221

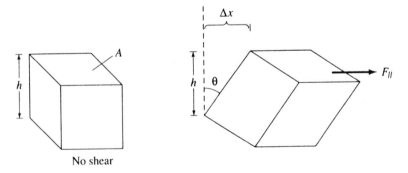

Figure 4.8-2
The shear force, $F_{\parallel}$, is parallel to the surface area, A, causing the surface to slip through an angle θ or the equivalent distance Δx.

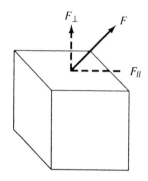

Figure 4.8-3
Force *F* produces both normal stress, $F_\perp$, and shear stress, $F_\parallel$.

• Strain is the deformation of a body that results from the application of stress. In other words, stress deforms a body and the actual deformation is called strain. Strain is the ratio of the change produced by the stress to the original length or volume. Therefore, strain is always dimensionless.

Elongation strain is associated with both tension and compression stresses. Both are the fractional change in length produced by the stress and given by the ratio of the change in length to the original length. Elongation strain is dimensionless:

$$\text{elong. strain} = \Delta\ell/\ell$$

Volume strain is the fractional change in volume produced by a stress. It is given by the ratio of the change in volume to the original volume:

$$\text{volume strain} = \Delta V/V$$

Note: The volume strain is also called the hydraulic strain. Tensile, compression, and shear stresses apply only to solids. The required restoring force is supplied by the fixed molecular structure of the solid. Fluids, however, cannot undergo these stresses, especially shear stress, because there is no fixed internal structure to supply the restoring force.

The only stress a fluid can undergo is called hydraulic stress. This kind of stress occurs when an external pressure is applied perpendicularly to the surface of the fluid.

Solids can also experience hydraulic stress as well as tensile, compression and shear stresses.

Shear strain is the ratio of the parallel displacement, Δx, to the length of the distorted side, *h*. (See Figure 4.8-2.).

$$\text{shear strain} = \Delta x/h$$

This direct proportion is shown as a straight line on the graph in figure 9-20 page 218.

• **Elastic moduli:** Stress and strain are directly proportional. The proportionality constant is called a modulus of elasticity:

$$\text{stress} = \text{modulus} \times \text{strain}$$

There are three main elastic moduli. All of them have the units of pressure, N/m^2.

Moduli are always reported as positive numbers.

Notice that the text uses the symbol (E) for Young's Modulus as shown in Eq. 9-5 page 220.

Young's modulus, Y, is the ratio of the tensile or compression stress to the corresponding strain:

$$Y = (F/A)/(\Delta\ell/\ell) = (F\ell)/(A\Delta\ell)$$

This can be rearranged to solve for the force, F:

$$F = (YA)(\Delta\ell/\ell) = k\Delta\ell$$

This equation has the form of Hooke's Law where k, called the force constant or the spring constant is a constant for a given body. k is inversely proportional to the length, ℓ, of the body and directly proportional to the product, YA:

$$k = YA/\ell$$

Y depends on the temperature as well as the identity of the substance. Most values of Y are on the order of 10^{10} N/m^2.

The **shear modulus,** S, is also called the rigidity modulus. It is the ratio of the shear stress, $F_\parallel/A$, to the shear strain, $\Delta x/h$:

$$S = (F_\parallel/A)/(\Delta x/h) = Fh/A\Delta x$$

which is exactly analogous to the Young's modulus equation.

The greater the value of S, the more the substance resists shear stress.

The **bulk modulus,** B, gives the ratio of stress to hydraulic strain. The stress is the pressure applied to the solid or fluid; the strain is the fractional change in volume:

Eq. 9-7 page 222

$$B = -P/(\Delta V/V) = -PV/\Delta V$$

The units of B are units of pressure, N/m^2. Since increasing the pressure always decreases the volume of the fluid or solid, the equation carries a negative sign to ensure that B is always a positive number.

B reflects how difficult it is to compress a substance. The more easily a substance is compressed, the smaller its value of B. For example:

$$B_{steel} = 16 \times 10^{10} \text{ N/m}^2$$
$$B_{air} = 1.01 \times 10^5 \text{ N/m}^2$$

The values of B are usually measured at an external pressure of one atmosphere.

B for most gases is constant at 1.01×10^5 N/m^2. Solids with large Young's moduli also tend to have large bulk moduli. The bulk moduli for liquids are not much lower than those of solids. Both generally have magnitudes of $\sim 10^{10}$ N/m^2. This suggests that liquids are not very compressible.

The **compressibility constant,** K, is the reciprocal of the bulk modulus and gives the fractional decrease in volume produced by a small increase in pressure:

$$K = 1/B$$

The units of K are square meters per newton, m^2/N.

Summary of Stress and Strain

Type of Stress	Stress	Strain	Elastic Modulus
Tension or Compression	$F_\perp/A$	$\Delta\ell/\ell_0$	$Y = (F_\perp/A)/(\Delta\ell/\ell_0)$
Hydrostatic Pressure	$P = F_\perp/A$	$\Delta V/V_0$	$B = -P/(\Delta V/V_0)$
Shear	$F_\parallel/A$	$\tan\theta$	$S = (F_\parallel/A)/\theta$

Shear strain = $\Delta x/h$ = tan $\theta \sim \theta$ if $\Delta x \ll h$. Although the angle is measured in radians, shear strain has no units because it is the ratio of two distances, x and h. Modulus = stress/strain.

C. Hydrostatics

Hydrostatics is the study of fluids at rest.

Any force acting on (or produced by) a static fluid must be perpendicular to the surface of the fluid component it acts on because any force parallel to the fluid surface will cause the fluid to flow. This changes the fluid from a hydrostatic to a hydrodynamic system.

Note: The surface of a fluid component need not occur at an interface between the fluid and some other substance. It includes the internal surfaces of any arbitrary volume element within the sample of fluid.

Static fluids cannot have parallel force components.

Since the frictional forces are always parallel to a surface involved, a static fluid has no static coefficient of friction.

Optional aside: Fluid lubricants are placed between two surfaces. Since the lubricant is at rest, there is no frictional force between it and either surface, which in turn reduces the friction between the two surfaces.

Sec. 10-4
page 241
• **Pascal's Principle** states that any pressure or change in pressure applied to an enclosed fluid at rest is transmitted uniformly and without any loss to all parts of the fluid and to the walls of the vessel containing the fluid. The pressure, as usual, acts perpendicularly to the surface of the fluid.

The pressure in a static fluid can be described either under the condition of negligible gravity or the condition where the effect of gravity cannot be neglected. Pascal's Principle generally refers to the former condition and can be restated as follows: If the effect of gravity is negligible, then the weight of the fluid is negligible and the pressure of the static fluid will be the same at all points in the fluid. This means that the pressure at the bottom of a tank of any fluid is the same as the atmospheric pressure applied at the top of the fluid.

The law of hydrostatic pressure is the application of Pascal's Principle to systems where the effect of gravity cannot be neglected. It states that the **hydrostatic pressure** of a fluid, P_h, at any given point is the sum of any external pressure, P_0, plus the weight of the fluid above that point:

$$P_h = F_{total}/A = (F_0 + F_g)/A = P_0 + F_g/A$$

where F_0 is the external force producing the external pressure, and F_g is the force due to gravity.

Sec. 10-2
page 238
gives details of
this derivation.
Since density, ρ, is mass over volume, m/v, and volume is area times height, Ah, then:

$$F_g = mg = \text{wgt} = \rho Ahg$$

where g is the acceleration due to gravity and wgt is the resulting weight. The contribution of the fluid to the hydrostatic pressure is the product of the density of the fluid, the height (or depth) at which the pressure is being evaluated, and g:

$$P = F_g/A = \rho hg$$

Since the external or atmospheric pressure is constant for a given sample, the change in pressure for any two points, A and B, in the sample depends only on their depths:

$$\Delta P = P_A - P_B = \rho g(h_A - h_B) = \rho g\Delta h$$

This means:

1. The pressure at the bottom of a tank of fluid must be greater than the pressure at the top.

2. Pressure depends only on the depth and is independent of the horizontal dimensions (area) of the surface. The pressure at any point in a static fluid is given by:

$$P = P_{external} + \rho g h$$

Therefore, if two surfaces are at the same depth in a fluid they will both experience the same pressure regardless of their areas. (The pressure on a square-inch sample of a given fluid is the same as the pressure on a square-mile sample of the fluid provided both samples are at the same depth.) See Figure 4.8-4.

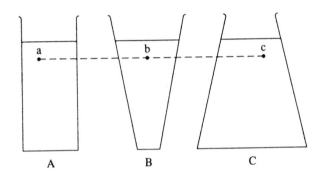

Figure 4.8-4
Three vessels with different shapes are filled with fluid to the same height. Points at the same height in the vessel will have the same pressure, $P_a = P_b = P_c$. Therefore, the pressure is the same at the bottom of the vessels despite their different shapes. However, the force exerted at levels of equal height in the different vessels will not be the same since the surface areas for the planes containing the points are different; $F_a \neq F_b \neq F_c$. The force exerted on the bottom of each vessel is different. Since force and area are directly proportional, the force is greatest for C and least for B.

Optional Aside: The atmosphere is a fluid that exerts a force on the surface of the earth. It produces atmospheric pressure which decreases with increasing altitude just as the pressure in a tank of water decreases from the bottom to the top.

There are several instruments that are used to measure the pressure in a static fluid. The two most common are the barometer and the manometer.

Sec. 10-5
page 242

The **barometer** is an evacuated closed-end cylinder that is completely filled with a fluid such as mercury and inverted with its open end under the surface of a reservoir of the same fluid. If the pressure due to the weight of the fluid, $\rho g h$, is less than the external pressure, P_0, of the atmosphere acting on the surface of the fluid in the reservoir, fluid will flow out of the cylinder and into the reservoir until $\rho g h = P_0$. The temperature at which the observations occur must be noted because density is temperature dependent.

A **manometer** can be used as a gauge to measure the pressure of a system. It consists of a U-shaped tube that is partially filled with a liquid of known density, often mercury or water. One end is attached to the system to be monitored, and the other is either left open to the atmosphere or closed and evacuated. In an open-end manometer, the level of the fluid in each arm of the U tube adjusts itself so that the sum of the pressure of the system, P_{sys}, and the fluid in one arm, $\rho g h_1$, equals the sum of the pressure of the atmosphere, P_0, and the fluid in the other arm, $\rho g h_2$. Thus the pressure of the system is:

$$P_{sys} = \rho g(h_2 - h_1) + P_0$$

In a closed-end manometer, $P_0 = 0$ because that side of the arm was evacuated. Therefore:

$$P_{sys} = \rho g(h_2 - h_1)$$

and depends only on the difference between the fluid levels in the two arms.

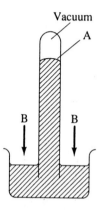

Figure 4.8-5
The fluid in a barometer rises under a vacuum until the pressure in the cylinder, A, equals the atmospheric pressure on the reservoir, B.

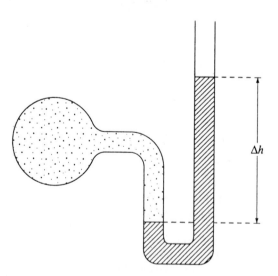

Figure 4.8-6
Illustration of an open-end manometer. In a closed-end manometer, $P_0 = 0$.

A hydraulic lever or lift is an example of the direct application of Pascal's Principle. A **hydraulic lift** consists of two pistons, one with a small surface area called the input

Figure 10-4
page 242 shows
an application of
Pascal's
Principle.

piston, and the other with a larger area called the output piston, connected by an enclosed fluid.

For a given pressure or change in pressure the larger the area subjected to that pressure, the greater the force produced:

$$\Delta P = P = F_{small}/A_{small} = F_{large}/A_{large}$$

A small force is applied to the piston with the smaller area. The resulting pressure is transmitted uniformly and undiminished throughout the fluid to the larger area piston where it exerts a proportionately larger force. The work done on the small piston must equal the work done by the large piston:

$$W = F_{input}d_{input} = F_{output}d_{output}$$

Since $F_{input} < F_{output}$ then $d_{input} > d_{output}$. The smaller piston moved a greater distance than the larger piston.

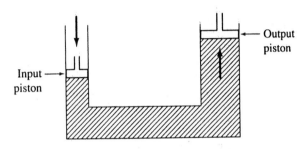

Figure 4.8-7

Hydraulic lift.

Sec. 10-6
page 244

"wgt" is weight.

• **Archimedes' Principle** states that a floating body is in static equilibrium and the upward or buoyant force, F_B, exerted by the fluid on a floating or on a submerged body equals the weight of the fluid displaced by the body. The body will sink until the weight of the fluid it displaces is equal to its own weight:

$$F_B = \rho vg = (m/v)(vg) = mg = wgt$$

The volume of the object that is below the surface of the fluid is equal to the volume of the displaced fluid, and the density is that of the fluid. Thus F_B is equal to the weight of the displaced fluid. Be careful: In order to float, the body must displace its own weight, not its own volume in fluid.

As a result of Archimedes' Principle:

See Ex. 10-3
page 245 and Ex.
10-4 for
examples of
problems using
Archimedes'
Principle.

1. If the density of the body is greater than the density of the fluid, $\rho_{body} > \rho_{fluid}$, then $wgt_{body} > F_B$ and the body will sink in the fluid (be completely submerged).

2. If the density of the body is less than the density of the fluid, $\rho_{body} < \rho_{fluid}$, then $wgt_{body} < F_B$ and the body will float on the surface of the fluid (be only partly submerged).

3. A body floating in a fluid appears to weigh less than it does in air because the buoyant force and the weight are antiparallel vectors. The weight apparently lost is equal to the weight of the fluid displaced by the body.

4. Since two masses cannot occupy the same space at the same time, the volume of a submerged body must equal the volume of the fluid it displaces.

The **center of buoyancy** is the point in a submerged or floating body where the total upward buoyant force can be considered to act. Since pressure increases with depth, the pressure at the top of a body is less than the pressure at the bottom of the body. Therefore, the center of buoyancy is not necessarily coincident with the center of mass (or gravity) for the body.

- Surface tension, γ, is the force per unit length perpendicular to the line of action of the force:

Sec. 10-13
page 261

$$\gamma = F_\perp / A$$

The SI units are newtons per meter, N/m.

It can be equivalently expressed as the potential energy per unit surface area:

The letter (A) refers to a length rather than an area. The letter used in the text is (L).

$$\gamma = E_{p(liquid)} / A$$

The SI units are joules per square meter, $J/m^2 = N/m$.

Liquid molecules are attracted to each other by cohesive forces and to molecules of other substances by adhesive forces.

The cohesive forces are great enough to produce definite surfaces or interfaces, but not so great that the liquid is rigid.

Internal molecules, those in the bulk of the liquid, are uniformly attracted in all directions by cohesive forces, and therefore feel no net force.

Molecules at interfaces are subject to both cohesive and adhesive forces which do not cancel. Therefore, surface molecules experience a net force or surface tension. This net force means that work must be done in forming the surface; therefore, the surface molecules have a potential energy which is directly related to the surface area.

At the air interface: Adhesive forces between the liquid and a gas (such as air) act upward away from the bulk liquid and tend to be much weaker than the liquid's cohesive forces, which act downward, toward the bulk liquid. The resulting net force causes the liquid to behave as if it had formed a surface membrane or skin.

In a vacuum: A volume of liquid falling in a vacuum forms a sphere because this shape provides the minimum surface area, and therefore the minimum potential energy, E_p, for the surface of a given volume. Liquid samples falling outside of a vacuum are distorted into nonspherical drops by gravity and air resistance.

At the glass interface: Adhesive forces between a liquid and a solid (such as the glass walls of a container) are usually greater than cohesive forces or the liquid-gas interactions and produce an effect known as capillary action.

Capillary action: A capillary or capillary tube is a thin, walled narrow diameter tube, usually but not necessarily made of glass. When it is partially inserted vertically into a reservoir of liquid, the liquid will rise in the tube because of the adhesive forces. In the capillary the surface molecules are subject to three forces:

1. The cohesive forces acting downward into the bulk of the liquid

2. The adhesive forces with the atmosphere which are antiparallel with the cohesive force (since this force is almost always much less than the cohesive force, it can be neglected or added to give a slightly smaller cohesive force)

3. The liquid-glass adhesive forces act towards the wall of the tube

If the capillary is narrow, the liquid will form a hemispherically curved surface called the meniscus.

For most liquids, including water, the adhesive force exceeds the cohesive force and the meniscus curves upward forming a concave surface. Liquid rises in the capillary to a level greater than that of the reservoir liquid. This process is called wetting.

A small number of liquids, most notably mercury, are nonwetting. Here the cohesive forces are greater than the adhesive forces. The meniscus curves down at the edges forming a convex outer surface. The liquid in the capillary is usually depressed below that of the reservoir.

The height to which the liquid is elevated by capillary action is inversely proportional to the diameter of the capillary and to the density of the liquid, and directly proportional to the surface tension.

For two samples of the same liquid (therefore, γ and ρ are constant) the smaller the radius of the capillary tube, the greater the elevation.

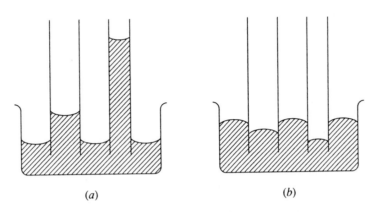

(a)   (b)

Figure 4.8-8
Meniscus for (a) a fluid where $F_{adhesive}$ is greater than $F_{cohesive}$ and (b) a fluid where $F_{adhesive}$ is less than $F_{cohesive}$.

Optional Aside: The upward force on the edges of the meniscus is the surface tension acting on the perimeter of the surface (where the liquid contacts the capillary wall):

$$F_{up} = \gamma \, (2\pi r)$$

The downward force is the weight of the liquid expressed as its volume $\pi r^2 h$ times its weight per unit volume, ρg:

$$F_{down} = (\pi r^2 h)(\rho g)$$

At equilibrium, $F_{up} = F_{down}$, and solving for the elevation, h, gives:

$$h = (2\gamma)/(\rho g r)$$

D. Hydrodynamics

Sec. 10-7
page 248

Hydrodynamics is the study of fluids in motion. It covers two broad categories of fluids, ideal fluids and real fluids.

- An **ideal fluid** is one that flows steadily around a body but exerts no force on the body. It is defined by four criteria:

   1. It is nonviscous. There is no internal frictional resistance to its flow.

   2. The density is constant; therefore, the fluid is incompressible.

   3. Its flow is a steady state flow. The magnitude and direction of the velocity at any given point is constant with respect to time.

4. The flow is irrotational. This means that the fluid will not cause a test body placed in it to rotate about an axis through the body's center of mass. However, the body can move in a circular path within the fluid.

figures 10-14,
10-15
page 248

- **Streamlines** are fixed paths that fluids appear to flow along. The molecules of the fluid move without rotational motion or turbulence. The direction of the velocity of the fluid at a given point is the tangent to the streamline at that point.

  1. The velocity at a given point in a streamline is constant.

  2. The velocity at different points in a streamline can be different.

  3. Streamlines can never cross. Since the velocity vector is tangent to the streamline, if two streamlines crossed it would imply that the fluid had two different velocities at that point.

  4. The magnitude of the velocity at a given point is proportional to the density of the streamlines per unit of cross-sectional area at that point. Decreasing the diameter of a tube increases the velocity of the incompressible fluid. This is represented by compressing the spacing between streamlines.

In **laminar flow** the fluid moves in layers between two surfaces in such a way that the resulting streamlines are parallel to the surfaces and to each other. These layers slide smoothly by each other. A layer is defined as the maximum thickness of a fluid where all parts have the same average velocity.

Optional Aside: Different layers of a fluid with laminar flow tend to have different average velocities. The layer of fluid adjacent to a surface adheres to it and tends to move with the same speed as the surface. This layer of fluid tends to drag the next layer along with it, and so on. The net result is that the velocity of subsequent layers decreases as you move away from a moving surface or towards a stationary surface.

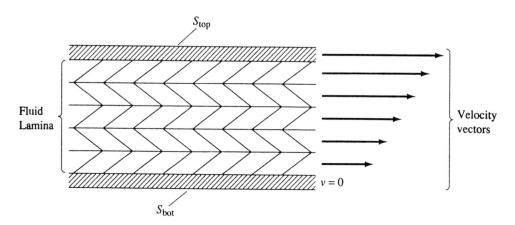

Figure 4.8-9
The top surface, S_{top}, is moving with a velocity v. The bottom surface, S_{bot}, is stationary. The fluid moves in layers between the two surfaces. The layer closest to S_{top} moves with a velocity approximately equal to that of the surface. Each successive layer moves with a diminishing velocity. The layer closest to S_{bot} has a velocity close to zero.

Ex. 10-9
page 250

- *Volume flow rate, Q:* The volume V of fluid flowing past a given area is the product of the cross-sectional area A, the velocity of the fluid v, and the time interval the flow is measured over, t:

$$V = Avt$$

The rate of flow Q is the volume per unit of time passing a given area and is given by the product of the velocity and the area crossed:

$$Q = V/t = Av$$

An SI unit of the volume rate is cubic meters per second, m^3/s.

figure 10-15
page 244

The **equation of continuity** states that for an incompressible fluid flowing through an enclosed tube, the volume flow rate Q is constant. For two points along the tube, $A_1v_1 = A_2v_2 =$ constant. Since the cross-sectional area of the tube and the velocity of the fluid are inversely proportional, increasing the diameter of a tube decreases the velocity of the fluid and vice versa.

Sec. 10-8
page 250

- **Bernoulli's equation** states that for an incompressible fluid with negligible viscosity moving with streamline flow, velocity and pressure are inversely related so that any increase in velocity of the fluid will produce a decrease in the associated pressure. It is an application of the Laws of Conservation of Energy and Mass; specifically it is the Work-Energy Theorem of moving fluids. It is used to describe the pressure associated with moving fluids when the viscosity is negligible, so that there is no loss of kinetic energy as thermal energy:

$$P + \rho gh + 1/2\ \rho v^2 = \text{constant}$$

See figure 10-19
page 252 for
examples of the
Bernoulli
principle.

Bernoulli's equation states that the sum of the three terms on the right-hand side is the same at every point in the fluid. Therefore, for any two arbitrary points:

$$P_1 + \rho gh_1 + 1/2\ \rho v^2_1 = P_2 + \rho gh_2 + 1/2\ \rho v^2_2 = \text{constant}$$

Where P is the pressure; ρgh is the potential energy per unit volume, E_p/V; $1/2\ \rho v^2$ is the kinetic energy per unit volume, E_k/V. All three units terms have units of pressure.

Case 1. Constant height: For horizontal flow, $h_1 = h_2$ and the equation reduces to:

$$P_1 + 1/2\ \rho v^2_1 = P_2 + 1/2\ \rho v^2_2$$

Pressure is greatest when the velocity (and thus the second term on each side) is least. The kinetic energy term can only be changed if work is done; therefore, there must be a net force acting on the fluid.

Case 2. Constant velocity: Here the equation reduces to:

$$P_1 + \rho gh_1 = P_2 + \rho gh_2$$

This is a manifestation of Pascal's Principle as observed in the operation of barometers and manometers.

Case 3. Constant pressure: The equation becomes:

$$\rho gh_1 + 1/2\ \rho v^2_1 = \rho gh_2 + 1/2\ \rho v^2_2$$

If two leaks are created at different heights in a water tank, the liquid exiting from the lower leak must have the greater velocity (bigger $1/2\ \rho v^2$ term) since it will have the smaller height (ρgh term).

- **Real fluids** are all fluids that do not meet the four criteria of ideal fluids. A real fluid does exert a force on any body moving through it. This frictional force arises from intermolecular interactions; the viscosity of the fluid produces a resisting force that is parallel to the surface of the body. This in turn causes the conversion and loss of some of the kinetic energy of the fluid motion as thermal energy. Thus, fluid near a body (surface) tends to move more slowly, forming a boundary layer. The flow and the loss of kinetic energy are directly proportional; if the flow is slow, the frictional loss of energy will be low.

Viscous force, F_v: A fluid flowing past a surface exerts a force, $F_\|$, that is parallel to the surface and in the same direction as the flow. In reaction, the surface exerts a force on the fluid called the viscous force, F_v, which is opposite to the direction of the flow. F_v represents the internal friction or internal inertia of a fluid that causes it to resist flowing. It is directly proportional to the velocity of the fluid, v, and to the area of the surface the fluid flows over, A, and inversely proportional to the distance between the fluid element and the surface, d. The proportionality constant is called the viscosity, η:

figure 10-24
page 255

$$-F_\| = F_v = \eta A v / d$$

Note: In laminar flow this internal friction produces a shear stress between adjacent layers.

Viscosity is the resistance of a fluid to flow. The greater the viscosity, the more difficult it is to get fluid motion. The difficulty with which a fluid flows is measured by the viscosity coefficient, η, which is temperature dependent and characteristic of the fluid. The SI unit of viscosity is the newton second per square meter, $N\,s/m^2$, which is also called the poiseuille, Pl. In the cgs system the unit is the poise, P:

$$1\ N\,s/m^2 = 1\ Pl = 10\ P$$

In laminar flow of fluids through pipes and tubes, the layers are arranged as concentric cylinders. The fluid velocity is greatest, v_{max}, for the central cylinder and is slowest for the outermost cylinder which is immediately adjacent to the walls of the pipe. The average velocity of different layers depends on the radial distance of the layer from the center of the pipe.

In turbulent flow there is no steady state streamline flow; the velocity at any given point is not constant but tends to change rapidly with time. Turbulence is the result of four factors: the density of the fluid, ρ, its average forward velocity, v, its viscosity, η, and the diameter of the pipe carrying the fluid, d.

The **Reynolds number**, $\mathcal{R}$, is a dimensionless quantity derived from the combination of the four factors above and used to predict the conditions of turbulent flow for a fluid system:

$$\mathcal{R} = \rho v d / \eta$$

1. If $\mathcal{R} < 2000$ the flow will be laminar.

2. If $\mathcal{R} > 3000$ the flow will be turbulent.

3. $\mathcal{R}$ values between 2000 and 3000 represent a transition region between laminar and turbulent flow patterns.

4. $\mathcal{R}$ is directly proportional to the density, velocity and tube diameter, and inversely proportional to the viscosity. Changing any combination of these appropriately can drive a system from laminar to turbulent flow or vice versa.

Practice Problems

"Problem Solving" page 243

1. A diver is swimming 10 meters below the surface of the water in a reservoir. There is no current, the air has a pressure of 1 atmosphere, and the density of the water is 1000 kilograms per cubic meter. What is the pressure experienced by the diver?

   A. 1.1 atm
   B. 1.99×10^5 Pa
   C. 11 atm
   D. 1.01×10^5 Pa

Eq. 9-5 page 220

2. The Young's modulus for steel is $2.0 \times 10^{11} N/m^2$. What is the stress experienced by a steel rod that is 100 centimeters long and 20 millimeters in diameter when it is stretched by a force of 6.3×10^3 N?

   A. 2.00×10^8 N/m^2
   B. 12.6×10^{12} N/m^2
   C. 3.15×10^8 N/m^2
   D. 4.0×10^{11} N/m^2

3. A 100-centimeter-long steel rod experiences a stress of 4.0×10^8 N/m^2 when it is stretched by a force of 10 N. The Young's modulus of steel is 2.0×10^{11} N/m^2. What is the strain on the rod?

   A. 5.0×10^3 m
   B. 5.0×10^3
   C. 2.0×10^{-3} m
   D. 2.0×10^{-3}

4. The aorta of a 70 kilogram man has a cross-sectional area of 3.0 square centimeters and carries blood with a velocity of 30 centimeters per second. What is the average volume flow rate?

   A. 10 cm/s
   B. 33 cm^3/s
   C. 10 cm^2/s
   D. 90 cm^3/s

"Problem Solving" page 243

5. A closed-end tube is evacuated and placed with its open end beneath the surface of a reservoir of mercury. The mercury rises to a height of 93 cm. If the density of mercury is 1.36×10^4 kg/m^3, what is the pressure at the bottom of the column of mercury?

   A. 1.24×10^5 Pa
   B. 1.24×10^5 atm
   C. 1.26×10^4 Pa
   D. 1.26×10^4 atm

6. At 20° C the density of water is 1 g/cm^3. What is the density of a body that has a mass of 100 grams in air and 25 grams in water?

Ex. 10-4 page 246

   A. 0.25 g/cm^3
   B. 0.75 g/cm^3
   C. 1.3 g/cm^3
   D. 4.0 g/cm^3

7. Brass has a density of 8.9 g/cm^3. A sample of brass is shaped into a perfect cube that has a mass of 71.2 grams. What is the length of each side of the cube?

   A. 2.0 cm
   B. 4.0 cm
   C. 8.0 cm
   D. 9.0 cm

8. What is the specific gravity of a bar of iron that has a mass of 192 grams and the dimensions 12 cm × 2 cm × 1 cm?

Sec. 10-1 page 238

   A. 8.0 g/cm^3
   B. 8.0
   C. 24 g/cm^3
   D. 24

9. Two insoluble bodies, A and B, appear to lose the same amount of weight when submerged in alcohol. Which statement is most applicable?

"Archimedes' Principle" page 245

   A. Both bodies have the same mass in air.
   B. Both bodies have the same volume.
   C. Both bodies have the same density.
   D. Both bodies have the same weight in air.

10. The bottom of each foot of an 80-kg man has an area of about 400 cm^2. What is the effect of his wearing snowshoes with an area of about 0.400 m^2?

    A. The pressure exerted on the snow becomes 10 times as great.
    B. The pressure exerted on the snow becomes 1/10 as great.
    C. The pressure exerted on the snow remains the same.
    D. The force exerted on the snow is 1/10 as great.

11. Two rectangular water tanks are filled to the same depth with water. Tank A has a bottom surface area of 2 m^2 and tank B has a bottom area of 4 m^2. Which statement about the

forces and pressures at the bottom of the tanks is correct?

A. Since F_A is less than F_B, then, P_A is less than P_B.

B. Since F_A is equal to F_B, then, P_A is less than P_B.

C. Since F_A is less than F_B, then, P_A is equal to P_B.

D. Since F_A is equal to F_B, then, P_A is equal to P_B.

"Pascal's Principle" page 241

12. In a hydraulic lift the surface of the input piston is 10 cm² and that of the output piston is 3000 cm². What is the work done if a 100 N force applied to the input piston raises the output piston by 2.0 meters?

A. 20 kJ
B. 30 kJ
C. 40 kJ
D. 60 kJ

Answers and Explanations

1. **B** The fluid is at rest (no currents) so this is a hydrostatic pressure calculation. In SI units 1 atm $= 1.01 \times 10^5$ Pa. Choice D can be eliminated because the pressure below the water must be greater than the pressure at the surface.

$$P_{diver} = P_{atm} + \rho g h$$
$$= (1.01 \times 10^5 \text{ Pa}) + (10^3 \text{ kg/m}^3)(9.8 \text{ m/s}^2)(10 \text{ m})$$
$$P_{diver} = 1.99 \times 10^5 \text{ Pa}$$

This problem is easily solved by estimating the answer: $9.8 \sim 10$; $1.01 \times 10^5 \sim 10^5$, therefore:
$$P_{diver} \sim (10^5 \text{ Pa}) + (10^3 \text{ kg/m}^3)(10 \text{ m/s}^2)(10 \text{ m})$$
$\sim 2 \times 10^5$ Pa which is closest to choice B.

2. **A** Stress is force per unit area, so neither the Young's modulus nor the length of the rod are needed to solve the problem. The area of the rod is $\pi r^2 = (3.15)(10 \times 10^{-3} \text{ m})^2$
$$= 3.15 \times 10^{-5} \text{ m}^2$$

Stress $= F/A$
$$= (6.30 \times 10^3 \text{ N})/(3.15 \times 10^{-5} \text{ m}^2)$$
$$= 2.00 \times 10^8 \text{ N/m}^2$$

3. **D** Strain is the ratio of change in length to the original length. Therefore start by finding the change in length:

$\Delta \ell = \ell(\text{Stress}/Y)$
$$= (1.00 \text{ m})(4.0 \times 10^8 \text{ N/m}^2)/(2.0 \times 10^{11} \text{ N/m}^2)$$
$$= 2 \times 10^{-3} \text{ m}$$

Strain is dimensionless, therefore choices A and C can be eliminated immediately.

Strain $= \Delta \ell/\ell = (F/A)/Y = \text{Stress}/Y$
$$= (4 \times 10^8 \text{ N/m}^2)/(2 \times 10^{11} \text{ N/m}^2)$$
$$= 2 \times 10^{-3} \text{ (no units)}$$

4. **D** $Q = Av = (3.0 \text{ cm}^2)(30 \text{ cm/s}) = 90 \text{ cm}^3/\text{s}$.

5. **A** The pressure at the bottom of the column is due only to the weight of the mercury in the tube (because it was evacuated).

$P = \rho g h$
$$= (1.36 \times 10^4 \text{ kg/m}^3)(9.8 \text{ m/s}^2)(93 \times 10^{-2} \text{m})$$

These values readily round off to:

$$P = (1.4 \times 10^4 \text{ kg/m}^3)(10 \text{ m/s}^2)(1 \text{ m})$$
$$\sim 1.4 \times 10^5 \text{ Pa}$$

Since all the terms were rounded off, the actual answer will be a little smaller than the estimate. Choice A is the numerical closest match that also has the correct units.

6. **C** Archimedes' Principle. The apparent loss of weight (or mass) of a submerged body equals the weight (or mass) of the fluid displaced.
   1. The mass of the displaced water is: 100 g − 25 g = 75 g
   2. The volume occupied by this mass of water is: 75 g/(1 g/cm³) = 75 cm³
   3. The volume of the body must equal the volume of the water displaced.
   4. The density of the body is:

$$\rho = m/V = 100 \text{ g}/75 \text{ cm}^3 = 1.3 \text{ g/cm}^3$$

7. **A** Volume $= m/\rho$, and $\text{side}_{cube} = V^{1/3}$
   This problem lends itself well to quick estimation; $8.9 \text{ g/cm}^3 \sim 9 \text{ g/cm}^3$; 71.2 g can be replaced by 72 which is the closest integer multiple of 9:

$$\text{side} = (m/\rho)^{1/3} \sim (72 \text{ g}/9 \text{ g cm}^{-3})^{1/3}$$
$$= (8 \text{ cm}^3)^{1/3} = 2 \text{ cm}$$

Note: 2 cm is also the value of the length of the side calculated when the exact values are used.

8. **B** Specific gravity of solids is the ratio of the density of the solid to the density of water, 1.0 g/cm^3. Since it is a ratio of identical units the specific gravity is dimensionless:

$$\rho = m/V = 192 \text{ g}/24 \text{ cm}^3 = 8.0 \text{ g/cm}^3$$

Therefore the specific gravity is simply 8.0.

9. **B** Archimedes' Principle. The weight lost is equal to the weight of the displaced fluid; therefore, both must have the same volume because the volume of the fluid equals the volume of the body.

10. **B** The force exerted by the man is his weight and it is assumed to be constant. This eliminates choice D. For a constant force, pressure and area are inversely proportional. The area of the snowshoes is ten times the area of the foot so that the pressure associated with the snowshoes is the inverse of 10 or 1/10 the pressure exerted by the foot.

11. **C** Hydrostatic pressure depends only on depth. Since both tanks are filled to the same depth, $P_A = P_B$. Since $P = F/A =$ constant, the tank with the larger bottom area will experience the greater force, $F_B > F_A$.

12. **D** The pressure produced by the input piston is:

$$P = F/A = 100 \text{ N}/0.001 \text{ m}^2$$
$$= 1.00 \times 10^5 \text{ N/m}^2$$

Pascal's Principle insures that this pressure is transmitted uniformly and undiminished throughout the fluid. This pressure produces a larger force at the larger area output piston:

$$F_{out} = PA = (1.00 \times 10^5 \text{ N/m}^2)(0.300 \text{ m}^2)$$
$$= 3.00 \times 10^4 \text{ N}$$

The work done is:

$$W = Fd = (3.00 \times 10^4 \text{ N})(2 \text{ m})$$
$$= 6.0 \times 10^4 \text{ N m} = 6.0 \times 10^4 \text{ J}$$
$$= 60 \text{ kJ}$$

Key Words

adhesion
amorphous solids
Archimedes' Principle
barometer
Bernoulli's equation
bulk modulus
capillary action
center of buoyancy
cohesion
compressibility constant
compression
crystalline solids

deformation
density
diffusion
elasticity
elastic limit
elongation strain
equation of continuity
hydrodynamics
hydrostatics
hydraulic lift
hydrostatic pressure
ideal fluid

laminar flow
manometer
moduli
Pascal's Principle
pressure
real fluids
Reynolds number
shear modulus
streamlines
volume strain
Young's modulus

4.9 General Wave Characteristics and Periodic Motion

Sec. 11-8
page 286

A. Waves: General Description

A wave is a general disturbance that is propagated (moves) through matter and/or space. This disturbance can:

1. change its magnitude from point to point along the line of propagation, and/or

2. change its magnitude at a given point with respect to time.

The common defining property of all waves is that wave motion is a mechanism that transfers energy from one point to another.

If the wave propagates through matter, this transfer of energy occurs without any net transfer of matter.

Waves carry energy and momentum obtained from the source of the wave. Intensity, I, is the power transported across a unit cross-sectional area; it is directly proportional to the square of the amplitude, $I \propto A^2$, and has the SI units of watts per square meter, W/m^2.

The transfer of energy can occur as a single event called a pulse, or as a series of consecutive displacements called a train of pulses. If a train of pulses has a regularly repeating pattern, it can be described by a sinusoidal wave.

- A **sinusoidal wave** is a repeating pattern that gives the relative displacement of a body from its equilibrium position. It is initiated by **simple harmonic motion** and can be described completely by a single wavelength and a single frequency. A repeating pattern can be described by a mathematical equation called a **periodic function.**

Note: **sine waves** and **cosine waves** are common sinusoidal waves.

The basic repeating pattern of any periodic function is called a cycle. One complete cycle of a sinusoidal wave has a value of 2π because it is equivalent to a body traveling once around a circle. A sinusoidal wave is completely described by the set of characteristics listed below:

The **amplitude,** A, is the maximum displacement from the equilibrium position. The equilibrium position has an energy value of zero. The transfer of energy by the wave is measured by the displacement away from equilibrium. The maximum displacement in the positive (maximum value of the function) direction defines a part of the curve called the crest of the wave. The maximum displacement in the negative direction (minimum value of the function) identifies the trough of the wave.

Note: By convention, in graphing sinusoidal waves the equilibrium position is parallel (or coincident) with the Cartesian x-axis and generally assigned the arbitrary value of zero. It also arbitrarily coincides with the direction in which the wave is traveling. Points above this line are considered positive and those below the line are considered negative.

The terms "nodes" and "antinodes" are used in the text only to describe properties of standing waves. see Sec. 11-12 page 299.

The points of zero displacement are called **nodes.** These mark the points where the wave crosses the equilibrium position. In sine waves the nodes occur at half wavelength intervals, $\lambda/2$.

The points of maximum displacement (positive or negative) are called **antinodes.** These correspond to the crests (maxima) and troughs (minima) of the wave. For sine waves there are two antinodes in each cycle (one positive and the other negative); antinodes occur half way between nodes.

The **phase,** Φ, of a sinusoidal wave is the relative position of its nodes (and antinodes) along the line of propagation with respect to some starting point or time.

Note: A single cycle of a sine wave starts and ends with a node and has a third node in the center of the cycle between the crest and trough. In contrast, a cosine wave starts and ends with antinodes and has an antinode in the center of the cycle. A node occurs between each pair of antinodes. A sine wave and a cosine wave are 90° out of phase with each other.

The **wavelength,** λ, is the distance between corresponding points on two successive cycles. It is generally measured from the crest of one wave to the crest of the adjacent wave, but any two corresponding points must give you the same value. The unit of wavelength is distance, which in the SI system is given in meters, m.

The letter "f" is used in the text for frequency rather than the Greek letter "ν".

Frequency, ν, is the number of complete cycles that pass a given point in one second. Since frequency is cycles per second, the unit of frequency is the reciprocal second, s^{-1}, also called the hertz, Hz.

The **period,** T, is the length of time required for the body to complete one cycle. The period is equal to the reciprocal of the frequency, ν:

$$T = 1/\nu$$

The unit of the period is the second, s; that of frequency is the reciprocal second or hertz, $s^{-1} = $ Hz.

Be careful not to confuse "ν" with the small letter "v" used for velocity.

The **velocity** (or speed), v, **of a sinusoidal wave** is equal to the product of the wavelength and the frequency:

$$v = \lambda\nu$$

The SI unit of speed and velocity is meters per second. The speed depends primarily on the medium the wave is traveling through. For the special case of **electromagnetic waves** traveling through a vacuum the speed is a constant, c:

$$v = c = 3.00 \times 10^8 \text{ m/s}$$

Sec. 11-10 page 291 gives a more detailed derivation of the energy in a wave.

The energy in a sine wave is the sum of the kinetic and potential energies:

$$E_{tot} = E_k + E_p = 1/2 \ mv^2 + 1/2 \ kx^2$$

where x is the displacement from the equilibrium (zero energy) position. At maximum displacement $x = A$, the amplitude, and $E_k = 0$, therefore:

$$E_{tot} = 1/2 \ kA^2$$

- *Types of waves:* There are two broad categories of waves, mechanical waves and electromagmetic waves. They are distinguished by the source generating the wave and the medium the wave propagates through.

Mechanical waves require a physical medium (gas, liquid, solid) for propagation; they cannot travel through a vacuum. The velocity of a mechanical wave depends on the strength of the source and the elastic properties of the medium.

The medium must be elastic so that the disturbance can displace particles of the medium from their original or equilibrium positions. Generally, the more elastic the medium, the more easily it is disturbed and restored to its equilibrium state.

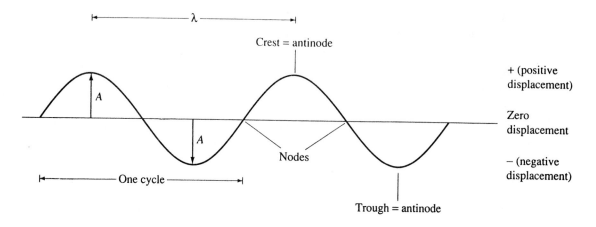

A = amplitude

Figure 4.9-1
Graph of a typical sine wave. The horizontal axis represents the value of the
equilibrium position of the body; it corresponds to zero displacement or zero
energy. The vertical axis gives the arbitrarily defined displacement of the body from
equilibrium. By convention, displacement above the horizontal axis is positive and
that below the horizontal is negative.

Sec. 11-9
page 289
Mechanical waves are most familiar and easiest to describe because they follow
Newton's laws of motion.
The **velocity of a mechanical wave** is:

$$v \, \alpha \, (\text{elasticity/inertia})^{1/2}$$

This velocity is:

1. Directly proportional to the square root of the elasticity of the medium. Elasticity
   is the "stiffness" of the bonds or interactions between the component particles of
   the medium. The greater the stiffness, the more easily (faster) the displacement
   will be propagated. The elastic property of the medium is given by the elastic
   moduli, Young's (Y), bulk (B), shear (S), or, if the medium is a string, by the Tension
   (T).

2. Inversely proportional to the square root of the inertia of the medium. The more
   massive the component particles of the medium, the slower the rate of propagation of a wave through the medium. Inertia can be measured by either the density
   (ρ) which is the mass per unit volume, or by the mass per unit length (μ).

Type of wave	Speed of wave
Sound wave in a fluid	$(B/\rho)^{1/2}$
Compressional or longitudinal waves in solids (P-waves)	$((B + 4/3 \, S)/\rho)^{1/2}$
Shear or transverse waves in solids (S-waves)	$(S/\rho)^{1/2}$
Compressional waves in a thin rod	$(Y/\rho)^{1/2}$
Transverse waves on a string	$(T/\mu)^{1/2}$

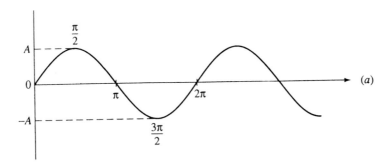

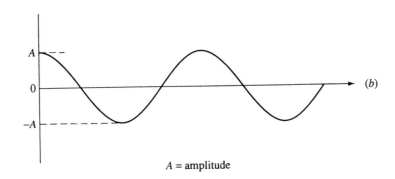

A = amplitude

Figure 4.9-2
Comparison of a single cycle of (a) a sine wave to that of (b) a cosine wave. Both are sinusoidal waves because of their shapes. They differ by a phase angle of 90° or one quarter of a cycle.

Electromagnetic waves, also called electromagnetic radiation, EMR, or light, do not require a physical medium for propagation; they can travel through a vacuum. EMR waves consist of two vector quantities, an electric field and a magnetic field which are mutually perpendicular. The strength of each of these fields varies sinusoidally and produces the disturbance (displacement) that changes with time and travels through space. (See Section 4.11).

1. In a vacuum all EMR waves travel with the same velocity, $c = 3 \times 10^8$ m/s.

2. In physical media the speed decreases. The decrease depends not only on the medium but also on the frequency of the light; the higher the frequency of the EMR, the greater its speed is affected by any physical medium. This leads to a phenomemon called dispersion and is the explanation for rainbows.

3. Electromagnetic waves generally arise from the interaction of mutually induced electric and magnetic fields that vary sinusoidally with time. However, moving matter also generates EMR waves called de Broglie matter waves (See Section 4.11.)

• Wave motion describes (1) the relation between the displacement and the direction of propagation and (2) the relative motion of the nodes (and antinodes) with time.

Note: In mechanical waves particles of the medium oscillate; in electromagnetic waves the electric and magnetic fields oscillate.

figure 11-20
page 289

In **transverse waves** the displacement (of particle position or field strength) is perpendicular to the direction of wave propagation. Electromagnetic radiation is always propagated as a transverse wave.

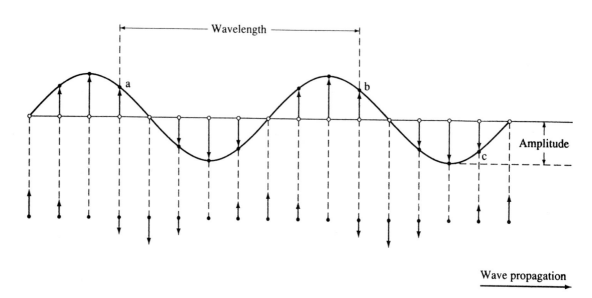

Figure 4.9-3
Transverse wave. The arrows are displacement vectors for the motion of particles of the medium in mechanical waves or the field strength in electromagnetic waves.

In **longitudinal waves** the displacement is parallel to the direction of propagation. Most sound waves and the compressional waves in a spring are longitudinal waves.

Note: Some wave motions are a combination of both longitudinal and transverse wave motions. The motion of water waves is such a combination because the water molecules move in circular or ovoid paths; the north-south (up-down) motion is the transverse component; the east-west (right-left) motion is the longitudinal component.

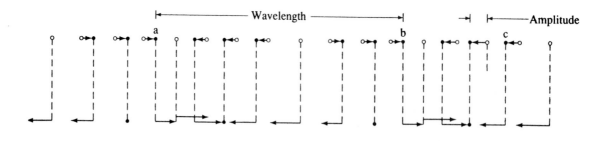

Figure 4.9-4
Longitudinal Wave. Arrows show displacement of particles of the medium.

In **traveling waves** the nodes move. This can be described in either of two equivalent remarks:

1. The position of a particular crest (or trough) will change with time.

2. The amplitude at a fixed point will change with time.

Water waves are an example of traveling waves.

In **standing waves** the positions of the nodes and antinodes are fixed and do not change with time. Colinear waves moving in opposite directions in the medium can add vectorally to produce a wave that appears to be stationary. A vibrating violin string is an example of a standing wave.

figure 11-29
page 296

- The **law of superposition** states that if two or more waves intersect at any point, the net displacement is the sum of the individual wave amplitudes (displacements) at that point. After the point of intersection the waves continue along their original paths unaltered by the interaction. Colinear waves interact at all points along the line of propagation.

   The result of superposition is called interference, and the effect observed depends on the phases and amplitudes of the waves at the points of intersection. See Figure 4.9-5 on the following page.

figure 11-31
page 29

1. **Constructive interference** occurs where the amplitudes have the same sign (both positive or both negative) at a given point.

   Perfect constructive interference occurs if the waves have the same phase and frequency so their antinodes coincide exactly (crests superimposed on crests and troughs superimposed on troughs). The amplitude of the resulting wave is always greater than the amplitude of any of the individual waves being superposed.

2. **Destructive interference** occurs where amplitudes have opposite signs at a given point.

   Perfect destructive interference occurs when the waves have the same frequency but are 180° out of phase with each other so that crests are superimposed on troughs. The amplitude of the resulting wave is always less than the amplitudes of any of the individual waves being superposed.

Note: For the special case where both waves have the same amplitude, perfect destructive interference produces complete destruction of the wave (zero amplitude at all points along the line of propagation).

Note: Fourier's theorem states that any wave (sinusoidal or not) can be reproduced by superposing various sine waves. Most real waves are not sinusoidal (i.e. having only one wavelength and one frequency) but are the result of the superposition of several waves.

B. Periodic Motion

In periodic motion a body repeats a certain motion, such as oscillating about its original equilibrium position. The body keeps traveling over the same definite path in equal intervals of time. Such a repeated path is called a closed path and each complete transit of the path is called a cycle.

In order to repeat the path, the body must regularly change its velocity, always altering the direction of the motion and often the speed as well. Therefore, the body must regularly experience a net force that provides the acceleration necessary to change the velocity.

Note: All periodic functions (patterns) can be replaced by a sine wave.

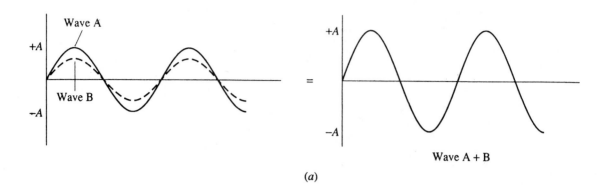

(a)

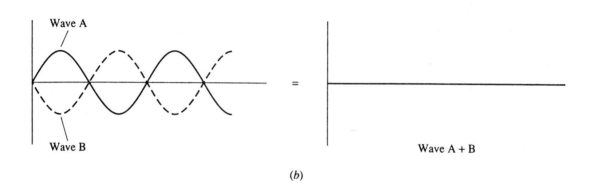

(b)

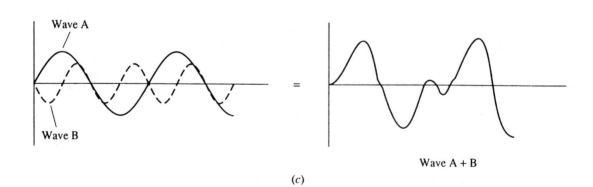

(c)

Figure 4.9-5
Interference patterns: (a) Perfect constructive interference. (b) Perfect destructive
interference. (c) Example of a real nonsinusoidal wave resulting from superposition
of two sinusoidal waves.

- Generating a sine wave from a periodic function: All periodic motion can be represented by a sine wave where one complete cycle of the wave corresponds to one complete transit of the path.

Sec. 11-4
page 278
Circular motion: The simplest way of generating a sine wave is by plotting the uniform (constant speed) circular motion of a body using a reference circle.

The graph of the motion is outlined in Figure 4.9-6. For uniform circular motion the length of time required for the body to complete one cycle of the circle is called the period and depends on the radius of the circle, R, and the speed of the body, v:

Period = Circumference of reference circle/speed

$$T = 2\pi R/v$$

A circle can be divided into four quarters or quadrants that are 90° apart. These are labeled Quads I, II, III and IV moving clockwise from point a.

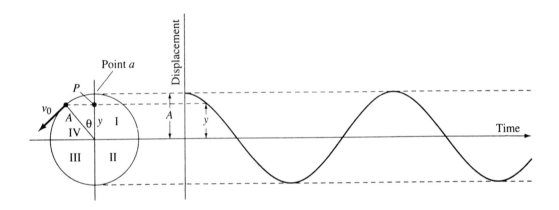

Figure 4.9-6
A body moves with constant speed in a circular path of radius A around some origin point, O. The horizontal axis of the sine wave graph is parallel to one of the diameters of the circle. The position of the circulating body relative to this diameter is plotted as the body moves around the circle from point a to point b to point c, etc, etc, until it returns to point a where it repeats the cycle. The radius to the current position of the body forms the phase angle θ with the radius to the original position of the body. The amplitude of the wave is the projection of the radius onto the y-axis (North-South pole line) of the circular motion.

Pendulum motion: An ideal pendulum consists of a dense mass called a bob suspended from a fixed point by a massless and nonstretchable string. The bob is free to swing back and forth about an axis. The axis is the equilibrium position that the pendulum has when it is at rest (zero velocity). See Figure 4.9-7.

Spring motion describes the motional effect of stretching or compressing a spring. A body attached to the spring will then oscillate about the rest or equilibrium position of the spring.

- Simple harmonic motion, SHM, occurs when a body moves back and forth around a definite path in equal intervals of time. SHM can always be described by a sine wave. SHM is a periodic motion where the force providing the acceleration is directly proportional to the displacement of the body from its equilibrium position and always directed towards the equilibrium position. Such a force is called a

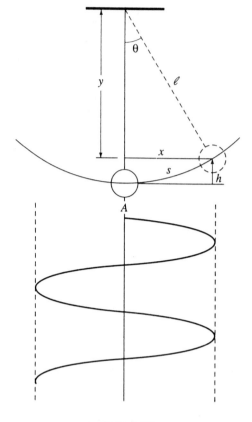

Notice that a
pendulum traces
out SHM only if
the angle is small
as shown in: Sec.
11-5 page 282.

Figure 4.9-7
The pendulum oscillates or vibrates around its equilibrium position; the bob is
displaced from the equilibrium position, $A = 0$, by an angle θ and during each
quarter cycle sweeps through an arc, S. The motion of the pendulum generates a
sine wave. The vertical axis is the zero displacement or equilibrium position;
displacements to the right are, arbitrarily, positive and those to the left, negative.

restoring force and is described by Hooke's Law. The motion may be linear or
angular.

Linear vibrational motion is typical of the oscillations of springs. Angular vibrational motion is typical of both circular motion and the motion of pendulums.
A body that moves with simple harmonic motion is called a harmonic oscillator.
Note: The general characteristics of SHM are those of waves. They are listed below
and will be modified later for specific types of SHM.

1. The period, T, is the time required for one complete cycle of the motion. It is
   independent of the amplitude of the motion.

2. The frequency, ν, is the reciprocal of the period:

$$\nu = 1/T = (k/m)^{1/2}/2\pi$$

Note: SHM is isochronous, which means the period and frequency are
independent of the amplitude of the motion. They depend only on the mass and
the force constant.

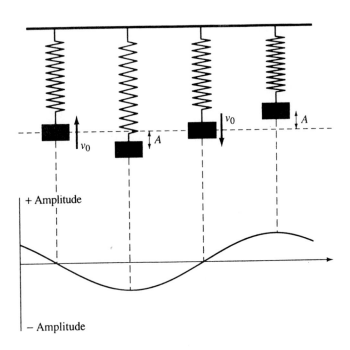

Figure 4.9-8
Graph of the motion of a body attached to an oscillating spring.

3. Displacement, d: There are two typical cases for measuring the displacement of a simple harmonic motion:

   Case 1: The body starts at its equilibrium position. That is, at time $t = 0$, the body has a displacement of $d = 0$ and is moving towards its maximum positive displacement (amplitude). Then, its displacement at any other time, t, is given by:

$$d = A \sin 2\pi\nu t = A \sin \omega t$$

   A is the amplitude of the maximum displacement, $\omega = 2\pi\nu$ is called the angular frequency of the motion and is used to express the frequency in radians per second instead of hertz.

   Case 2: The body starts at its maximum displacement position. That is, at time $t = 0$, the body has a displacement of $d = A$ and is moving towards its equilibrium position (where $d = 0$). Then, its displacement at any other time t is given by:

$$d = A \cos 2\pi\nu t = A \cos \omega t$$

4. The velocity, v, of the body at any time is given by:

$$v = 2\pi(A^2 - d^2)^{1/2}$$

   where the value of t determines the value of d. There are two roots to the square root equation: $+v$ indicates the body is moving towards its maximum positive displacement (crest of the wave); $-v$ indicates the body is moving towards its maximum negative displacement (trough of the wave).

5. The acceleration, a, at any time t is given by:

$$a = -4\pi^2\nu^2 d = -(k/m)d$$

   The acceleration has its maximum value at the points of maximum displacement (positive or negative); the acceleration is zero at the equilibrium position ($d = 0$).

   Note: At the turning points where the acceleration has its maximum values the velocity is zero because all of the energy is potential. At the equilibrium point where the acceleration is zero the velocity has its maximum value because all of the energy is kinetic.

Sec. 11-5
page 282

• *Motion of a pendulum*

Characteristics of the wave function:

1. Period: The equation for the period of a SHM is modified for the specific case of an ideal pendulum. The displacement of the pendulum is directly proportional to its length, ℓ, and the acceleration is due only to gravity, g, therefore:

$$T = 2\pi(\ell/g)^{1/2}$$

2. Frequency, ν, is the reciprocal of the period; therefore:

$$1/T = \nu = (1/2\pi)(g/\ell)^{1/2}$$

Note: The period and the frequency are independent of the mass of the pendulum.

Kinetic and potential energy of the motion: At all points in the path the energy is the sum of the potential and kinetic energy terms.

1. At the points of maximum displacement the bob of the pendulum has only potential energy, E_p = maximum, and no kinetic energy, $E_k = 0$. The velocity at these points is zero.

2. At the equilibrium position, the bob has only kinetic energy, E_k = maximum, and no potential energy, $E_p = 0$. The velocity of the bob is maximal at this point.

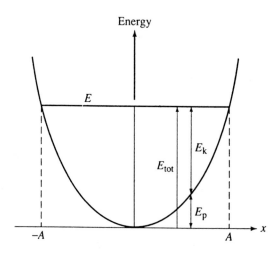

Figure 4.9-9
Distribution of E_p and E_k during the swing of a pendulum.

Sec. 11-1
page 274

• *Motion of a spring*

Motion of a spring is vibrational SHM motion described by Hooke's Law.

Hooke's Law was empirically derived from observing the motion of spring systems. It states that when an elastic body such as a spring is deformed, its displacement x is proportional to the applied force. The proportionality constant is called the force constant, or **spring constant,** k, and depends on the "stiffness" of the spring. The larger the value of k, the more difficult it is to stretch or compress the spring. The SI unit for k is the newton per meter, N/m or kg/s².

When the spring is deformed it exerts a restoring force, F_r, that is always oriented towards the equilibrium position, which means it is always antiparallel to the original deforming force:

$$F_r = -kx$$

Characteristics of the motion: There are two general descriptions, the case of a horizontal spring system and the case of a vertical spring system.

figure 11-2
page 275

Case 1: A horizontal spring system consists of a spring attached to a mass at one end and to a vertical wall at the other.

1. At equilibrium, the distance of the mass from the wall is the rest or equilibrium length of the spring. If an applied force stretches the spring, the mass is displaced from its equilibrium position by some distance A (which equals the maximum amplitude of the resulting motion). Because of the "stiffness" of the spring, a restoring force develops that opposes the applied force. The restoring force is at a maximum at this point.

2. When the applied force is removed, the restoring force accelerates the body back towards the equilibrium position. As the body approaches the rest position, the magnitude of F_r steadily decreases.

3. However, the body has momentum, *mv*, upon reaching the equilibrium position and tends to "overshoot" it.

4. As the body passes the rest point the spring is compressed, producing a restoring force in the opposite direction.

5. The restoring force is maximal at the turning point, which is the point of maximum compression.

6. Step 2 occurs in the opposite direction, accelerating the body back towards the rest position.

7. The body overshoots the rest point.

8. The cycle repeats.

figure 11-3
page 275

Case 2: A vertical spring system consists of a spring attaching a body to a horizontal beam or ceiling. The argument describing this motion is identical to that of case 1. The only difference is that the spring has two equilibrium lengths. The unloaded spring will have a particular equilibrium length. Loading the spring with a body produces a second equilibrium length that depends on the stiffness of the spring and on the mass of the body. The spring-mass system is then displaced from and oscillates about this latter equilibrium point.

Energy: At the turning points in the motion (maximum stretch and maximum compression) the velocity decreases to zero in prelude to reversing direction. All of the energy is potential because the kinetic energy is zero. As the body passes through the equilibrium position, the potential energy drops to zero and all of the energy is now kinetic. At positions between the rest and turning points the energy is the sum of the kinetic and potential energy contributions.

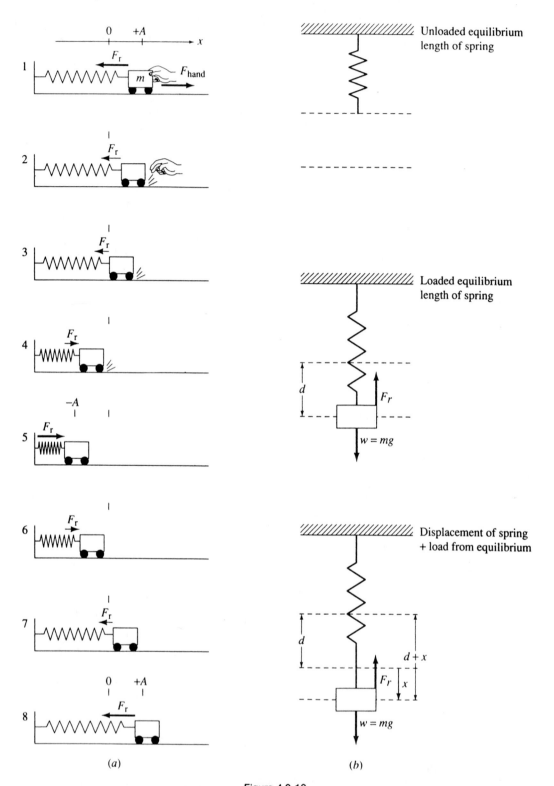

Figure 4.9-10
(a) The 8 steps describing the motion of a horizontal spring-mass system. (b) Motion of a vertical spring-mass system.

Practice Problems

1. A simple pendulum has a period of 4.63 seconds at a place on the earth where the acceleration of gravity is 9.82 m/s². At a different location the period increases to 4.64 seconds. What is the value of g at this second point?

A.   9.78 m/s²
B.   9.82 m/s²
C.   9.86 m/s²
D.   Cannot be determined without knowing the length of the pendulum

Eq. 11-9 page 288 is true for all sinusoidul waves.

2. What is the wavelength of a transverse wave with a velocity of 15 meters per second and a frequency of 5.0 hertz?

A.   3.0 m
B.   10 m
C.   20 m
D.   45 m

3. What is the optimum difference in phase for maximum destructive interference between two waves of the same frequency?

A.   360°
B.   270°
C.   180°
D.   90°

4. Standing waves can be formed if coincident waves

A.   have the same direction of propagation.
B.   have the same frequency.
C.   have different amplitudes.
D.   have different wavelengths.

Problems 5, 6, nd 7 all involve the relationship between the period and the length of a pendulum. Eq. 11-8 page 283.

5. A pendulum with a length ℓ has a period of 2 seconds. In order for the pendulum to have a period of 4 seconds, we must

A.   halve the length.
B.   quarter the length.
C.   double the length.
D.   quadruple the length.

6. If a pendulum 12 meters long has a frequency of 0.25 hertz, what will be the period of a second pendulum at the same location if its length is 3.0 meters?

A.   2.0 s
B.   3.0 s
C.   4.0 s
D.   6.0 s

7. A pendulum clock is losing time. How should the pendulum be adjusted?

A.   The weight of the bob should be decreased so it can move faster.
B.   The length of the wire holding the bob should be shortened.
C.   The amplitude of the swing should be reduced so the path covered is shorter.
D.   None of the above.

8. A 20-kg weight is attached to a wall by a spring. A 5.0 Newton force horizontally displaces it 1.0 meters from its equilibrium position along a frictionless floor. What is the closest estimate of the period of the oscillation of the weight?

A.   2.0 seconds
B.   6.0 seconds
C.   13 seconds
D.   16 seconds

Answers and Explanations

1. **A**   The answer can be determined easily without doing a numerical solution. Rearrange $T = 2\pi(\ell/g)^{1/2}$ to $Tg^{1/2} = $ constant. The period is inversely related to the square root of the acceleration due to gravity. Since T has increased, both $g^{1/2}$ and g must decrease. Choice A is the only value of g that is less than the original 9.82 m/s².

$$g_2 = (T_1^2 \dot{g}_1/T_2^2)$$
$$= (4.63 \text{ s})^2(9.82 \text{ m/s}^2)/(4.64 \text{ s})^2$$
$$= 9.78 \text{ m/s}^2$$

2. **A**   Wavelength is velocity divided by frequency. The formula is invariant to the type of wave involved.

$$\lambda = v/\nu = (15 \text{ m/s})/5.0 \text{ s}^{-1} = 3.0 \text{ m}$$

3. **C**   Two waves are completely out of phase when their antinodes coincide so that each crest on one wave coincides with a trough on the other. This occurs when the waves differ by 180°.

4. **B**   In standing waves the nodes are stationary. This can be accomplished when two waves with the same frequency travel in opposite directions.

5. **D** In a pendulum the period and the square root of the length are directly proportional:

$$T = 2\pi(\ell/g)^{1/2} \text{ so that } T/\ell^{1/2} = \text{constant}$$

To double the period you must double the square root of the length. To double the square root of the length you must quadruple the length:

$$(4\ell)^{1/2} = 4^{1/2}\ell^{1/2} = 2\ell^{1/2}$$

6. **A** The frequency is the reciprocal of the period: $\nu = 1/T$, so that $T = 4.0$ second. Period and the square root of the length are directly proportional. Therefore the ratio of the two periods is:

$$T_1/T_2 = (\ell_1/\ell_2)^{1/2} = (12/3)^{1/2} = 4^{1/2}$$
$$= 2.0 \text{ seconds}$$

7. **B** The period of a pendulum is directly related to the square root of the length of the cord holding the bob. It is independent of the weight and amplitude.

8. **C** The force constant is: $k = |F|/|\Delta x| = 5.0 \text{ N}/1.0 \text{ m} = 5.0 \text{ N m}^{-1}$. The period is:

$$\begin{aligned}
T &= 2\pi(mk^{-1})^{1/2} \\
&= 2(\pi)(20 \text{ kg}/5.0 \text{ N m}^{-1})^{1/2} \\
&= 2\pi(2)(\text{kg m N}^{-1})^{1/2} \\
&= 4\pi(\text{kg m N}^{-1})^{1/2} \sim 4(3)(\text{kg m N}^{-1})^{1/2} \\
&= 12(\text{s}^2)^{1/2} = 12 \text{ s}
\end{aligned}$$

Note: $N = \text{kg m s}^{-2}$, therefore $\text{kg/m N}^{-1} = \text{kg m/N} = \text{kg m s}^2/\text{kg m} = \text{s}^2$

Key Words

amplitude	law of superposition	sine waves
antinodes	(of waves)	sinusoidal waves
circular motion	longitudinal waves	spring constant
constructive interference	mechanical waves	spring motion
cosine waves	nodes	standing waves
destructive interference	pendulum	transverse waves
electromagnetic waves	pendulum motion	traveling waves
energy of waves	period	velocity of a mechanical wave
frequency	periodic function	velocity of a sinusoidal wave
gamma rays	phase	wavelength
Hooke's Law	simple harmonic motion	

4.10 Mechanical Waves: Acoustic Phenomena

Sec. 12-1
page 308

A. Sound Waves

Sound waves are the transmission of mechanical energy produced when a source initiates a disturbance in an elastic medium. Sound does not travel through a vacuum. Simple sound waves are sinusoidal with well defined wavelengths, frequencies and amplitudes.

- The **sonic spectrum** is the range of frequencies over which "sound waves" can be propagated.

The upper limit of the spectrum is well defined. It has a minimum frequency of 10^9 Hz for gases. The value for solids and liquids is higher. The frequency of the sound depends inversely on the interparticle distance of the medium. In solids and liquids, these distances are smaller than in gases so the frequencies are higher.

The lower limit is not well defined. However, earthquakes are common low-frequency phenomena producing wavelengths in the kilometer range and corresponding frequencies in the region of 10^{-1} Hz.

- The **audio range** is the region of the sonic spectrum that can be perceived by the human ear. For the average person it covers frequencies from 20 Hz to 20,000 Hz. Frequencies above this range are called ultrasonic frequencies, and those below this range are called infrasonic frequencies.

Optional aside: Do not confuse sound waves (which are mechanical waves) with radio or television waves (which are electromagnetic waves).

figure 11-22
page 290

- In fluids sound is transmitted only as longitudinal waves. The particles of the medium oscillate parallel to the direction of propagation of the wave. The velocity of propagation depends on the temperature, pressure and identity of the fluid.

The relative velocity is:

$$v = (B/\rho)^{1/2}$$

where B is the bulk modulus of the fluid, and ρ is its density. B is the ratio of stress to strain in the fluid. The stress is the change in pressure, and the strain is the fractional change in volume that the stress produces:

$$B = \Delta P/(\Delta V/V)$$

The larger the magnitude of B the more difficult it is to compress the fluid.

Optional aside: This is a check of the units in the equation for the velocity:

$$B/\rho = (N/m^2)/(kg/m^3) = [(kg\ m/s^2)/m^2/(kg/m^3) = (kg/s^2\ m)/(kg/m^3) = m^2/s^2$$

$$v = (B/\rho)^{1/2} = (m^2/s^2)^{1/2} = m/s$$

The velocity of sound in air at 0°C and one atmosphere of pressure is 330 m/s or ~ 740 miles per hour. The velocity in water under the same conditions is about four times as great, ~ 1402 m/s.

Optional aside: The density of water is ~ 1000 times that of air, so we might expect sound to travel faster in air. However, $B_{water} > B_{air}$ because water is less compressible than air, therefore, $\Delta V/V$ is less for water than air for the change in pressure, ΔP. That is, the larger bulk modulus of water causes sound to have a greater velocity in the water.

"Speed at sound
in air"
page 309
The velocity of sound in fluids increases with increasing temperature at the rate of approximately 0.6 m/s per 1°C, or ~ 2 ft/s °C. In air all sound waves have the same velocity at a given temperature regardless of their frequencies.

Note: Temperature affects the velocity of sound in all media. However, the effect is greatest for gases and least for solids.

- In solids sound can be transmitted as longitudinal waves and as transverse waves. Transverse waves are possible because of the three-dimensional lattice structure of solids. This structure allows particles to move perpendicularly to the line of propagation. The restoring force for this perpendicular displacement is weak; therefore, for a given frequency the velocity of the transverse wave will be less than the velocity of the corresponding longitudinal wave.

Optional aside: The distance from an observer to the source of an earthquake can be determined by measuring the difference in the arrival times of the transverse and the longitudinal waves generated by the source.

In solids the velocity of a longitudinal wave is given by:

$$v = (Y/\rho)^{1/2}$$

where Y is the Young's modulus, which is analogous to the bulk modulus, B. The distinction between Y and B is that fluids expand uniformly while solid samples tend to change length but not diameter.

The larger the value of Y, the more difficult it is to compress the solid and the smaller the period of oscillation of the particles and, therefore, the greater the velocity. The velocity of sound in solids is about 15 times that of air or ~ 5×10^3 m/s.

In solids the velocity of transverse waves is given by:

$$v = (T/\mu)^{1/2}$$

T is the tension or stiffness of the medium. μ is the mass per unit length of the medium.

B. Characteristics of Sound

The characteristics of sound can be described by three physical properties: intensity, frequency, and harmonic composition. Physical properties are objective because they are quantitative and can be measured independently of the observer.

Each physical property of sound has a subjective counterpart: loudness, pitch, and timbre respectively. Subjective properties are qualitative and cannot be measured directly; their values depend on the perceptions of the observer.

figure 11-24
page 292
- **Intensity,** I, is the average rate per unit area (normal to the direction of propagation) at which sound energy is transferred by the sound wave. It is the ratio of power per unit area:

$$I = Power/A$$

and has the units of watts per square meter, W/m^2.

Intensity is more important than total energy carried by the wave because the wave spreads over larger areas as it moves away from its source so that I drops with increasing distance from the source, as shown in Figure 4.10-1.

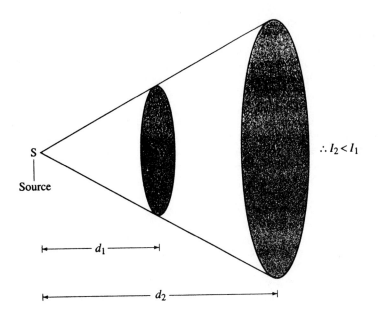

$$\therefore I_2 < I_1$$

Figure 4.10-1

Intensity is inversely proportional to the square of the distance from its source.

For a given frequency v, the intensity I, is directly proportional to the square of the amplitude a of the sound wave.

The **threshold of hearing,** I_0, is the minimum intensity of sound that is audible to the average listener. It has a value of $\sim 10^{-12}$ W/m^2 for sound with a frequency of $\sim 10^3$ Hz.

Sound level, β: The intensity of a sound I is measured with respect to the threshold of hearing. This ratio is called the sound level:

$$\beta = 10 \log I/I_0$$

See Sec. 12-2 page 310 for more details on the definition of the decibel.

1. The dimensionless ratio I/I_0 is called the **decibel,** dB. The range of perceptible intensities runs from a low of $I_0 \sim 10^{-12}$ W/m^2 to a high of ~ 1 W/m^2. This range covers 12 orders of magnitude; therefore, intensity is measured using the logarithm scale.

2. The threshold intensity, I_0, has a sound level value of $\beta = 0$ dB.

3. The maximum intensity, which marks the onset of pain, has a sound level of $\beta = 120$ dB.

4. The relationship between sound levels and intensities:

β (dB)	I/I_0
0	$10^0 = 1$
10	$10^1 = 10$
20	$10^2 = 100$
30	$10^3 = 1000$
.	.
.	.
120	10^{12}

Ex. 12-2 page 311 shows a typical calculation using the sound level relationship.

5. The difference in β is constant for each change in order of magnitude of intensity. That is, β changes by 10 dB in going from an intensity of 10^0 to 10^1 and in going

Ex. 12-4
page 313 shows a
typical
calculation
involving both
the intensity and
the sound level.

from 10^1 to 10^2, etc, so that for a given frequency, there is a uniform sensation to the changes in intensity.

Loudness is an auditory sensation and depends on the perceptions of the listener. It is the listener's subjective perception of sound level.

Figure 12-6
page 315 shows
this relationship.

Optional aside: The distinction between intensity and loudness occurs because the human ear is not equally sensitive to all frequencies. Although waves with greater intensity are generally louder, high-frequency waves do not seem as loud as lower-frequency waves of the same intensity. For example, a 20 dB sound with a frequency of 1000 Hz sounds louder than a 20 dB sound with a frequency of 100 Hz even though both have the same intensity.

- **Frequency,** ν, is a physical property that expresses the number of waves (cycles) per unit of time passing a given point.

The text refers to
the sound level as
the intensity
level.

Pitch is the subjective perception of frequency and depends on the frequency the ear receives.

1. Two waves with frequencies in a ratio of 2:1 are an octave apart.

2. Major chords in music are 4 frequencies in the ratio of 4:5:6:8.

3. At frequencies greater than 3000 Hz, pitch increases with intensity of a given frequency. Increasing the intensity of a high-frequency sound makes its frequency seem even higher than it is.

4. At frequencies below 2000 Hz, pitch decreases with increasing intensity of a given frequency, so a sound seems lower as it becomes more intense.

Sec. 12-5
page 316

- **Harmonic components:** The lowest frequency produced by a sound source is called its fundamental frequency or first harmonic. Harmonics that are whole number multiples of the fundamental are called overtones. The second harmonic is the first overtone. It lies an octave above the fundamental and vibrates with twice the frequency of the fundamental. The third harmonic is called the second overtone. It is two octaves above the fundamental and vibrates at three times the fundamental frequency.

Sec. 12-6
page 320

Timbre or quality, is the subjective perception of the harmonics. The number and relative intensities of harmonics present in a sound depends on the source, such as a musical instrument, producing the tones. This different combination of harmonics is what allows us to distinguish tones of the same frequency produced by different instruments.

C. Properties of Sound

Sound waves exhibit all the typical properties of mechanical waves.

- **Reflection:** Waves are reflected whenever they come to a boundary or change in the medium they are traveling through. The boundary may be obvious, such as the physical barrier of a wall; or less obvious, such as the change in the density of air produced by local changes in temperature.

For mechanical waves, the phase of the reflected wave depends on whether the boundary is "fixed" or "free."

If the boundary is fixed, the reflected wave is 180° out of phase with the initial wave. A fixed boundary is one that is fairly inelastic and does not permit the wave to be transferred through it. Most sound waves are reflected this way because they usually hit solid boundaries such as walls.

Optional aside: At the interface between the medium transmitting the wave and the boundary, in agreement with Newton's third law, the force of the wave on the wall is countered by an equal and opposite force produced by the wall on the medium. If the pulse of the wave first reaching the boundary exerts an upward force on the wall, the wall will exert a downward force on the medium and the resulting reflected wave will be exactly the reverse of the incident wave.

If the boundary is free, (not fixed), the reflected wave will be in phase with the initial incident wave.

An **echo** is a single reflection of a sound wave. Multiple reflections of a sound wave are called reverberations.

Optional aside: The human ear can retain awareness of sounds for $\sim$ 0.1 second, therefore the ear cannot perceive all echoes or reverberations.

1. If the reflection time is greater than 0.1 second, the ear will hear the echo or reverberation.

2. If the reflection time is less than 0.1 second, the echo will not be perceived because it will be masked by the memory of the original sound.

3. For perceptible echoes to occur, the reflecting surface must be more than 16.5 meters away so that the sound will have to travel a 33-meter round trip, which takes $\sim$ 0.1 second.

figure 11-32
figure 11-33
page 298

• **Diffraction:** Sound waves can be diffracted, or bent around the corners of obstacles, so that the wave is propagated on the other side of the obstacle and not just absorbed or reflected by the obstacle. The amount of diffraction is a function of the wavelength of the sound and the slit width of the obstacle.

If the wavelength is $<<$ slit, no diffraction will occur. The sound wave will pass through the slit unaltered.

If the wavelength is $\geq$ slit, diffraction occurs. Therefore, long-wavelength sounds (low-frequency sounds), can be heard farther and more clearly than short-wavelength (high-frequency) sounds.

Sec. 12-8
page 324

• **Doppler effect** is the change in frequency or pitch produced when the source of the wave or the detector of the wave or both are in motion with respect to the medium transmitting the wave. The greater the relative speed, the greater the shift in frequency. The shift is given by:

$$\nu' = \nu[(v + v_D)/(v - v_S)]$$

where ν' is the perceived frequency and ν is the actual frequency. v is the velocity of the wave in the medium, v_D is the velocity of the detector and v_S is the velocity of the source producing the sound.

If the detector is stationary and the source is moving, the equation becomes:

$$\nu' = \nu[v/(v - v_S)]$$

If the source is moving towards the detector, the denominator becomes $v - v_S$ so $\nu' > \nu$ and the pitch sounds higher.

If the source is moving away from the detector, the denominator becomes $v + v_S$ so $\nu' < \nu$ and the pitch sounds lower.

If the source is stationary and the detector is moving, the equation becomes:

$$\nu' = \nu(v + v_D)/v$$

If the detector is moving towards the source, the numerator becomes $v + v_D$ so $\nu' > \nu$ and the pitch sounds higher.

If the detector is moving away from the source, the numerator becomes $v - v_D$ so $\nu' < \nu$ and the pitch sounds lower.

If both the source and the detector are moving fairly slowly, the equation can be approximated by:

$$\nu' \sim \nu[1 \pm (v_R/v)]$$

where v_R is the relative speed between the source and the detector and is given by:

$$v_R = |v_S \pm v_D|$$

Note: Although the Doppler Effect was described for sound waves, the description and equations given apply to any wave, mechanical or electromagnetic.

See figure 12-19 page 324 for a different way of showing the two situations.

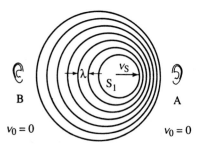

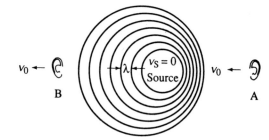

The source moves toward observer A and away from observer B at speed v_S

Observer A moves toward the stationary source at speed v_0, while observer B moves away from the source at the same speed

Figure 4.10-2

The Doppler Effect.

"Beats"
page 323

• **Beats** are the interference pattern produced by superposing two waves of slightly different frequencies being propagated in the same direction. The amplitude of the resulting wave varies with time.

For two waves, A and B, with frequencies ν_A, ν_B, and periods T_A, T_B, if $\nu_A > \nu_B$ then $T_A < T_B$, and the beat frequency is:

$$\nu_{beat} = \nu_A - \nu_B = 1/T_A - 1/T_B$$

because $\nu = 1/T$.

The human ear can detect beats with a frequency of up to 6–7 Hz.

Frequencies that differ by less than 2 or 3 Hz are not perceived as beats. They are perceived as being out of tune.

At higher frequencies it is more difficult to distinguish beats and the two frequencies merge to give the listener an average frequency:

$$\nu_{average} = 1/2(\nu_A + \nu_B)$$

This average frequency is either perceived as consonant (pleasing to the ear) or dissonant (displeasing to the ear), depending on the ratio of the two frequencies.

Note: Although beats were described for sound waves the description and equations given apply to any wave, mechanical or electromagnetic.

figure 12-13
page 323

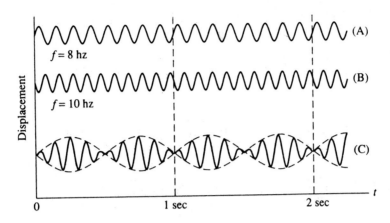

Figure 4.10-3

Beats.

D. Music

Initial and reflected waves superpose to produce standing waves in a medium with stationary nodes. The deliberate production of combinations and sequences of these standing waves that are pleasant to the ear is called music and is very subjective. Most musical instruments can be divided into two broad categories, strings and pipes.

Strings are instruments like guitars, violins, pianos, and the like, where the source of the sound is a vibrating string that produces a compression-rarification wave in the air that is detected by the ear as sound.

The frequency of a wave produced by a string varies with the string's length, diameter, tension and density. The relationship of frequency to these variables are summarized in the four **laws of strings:**

Ex. 12-5
page 316 shows
how one can use
this relation to
solve a problem.

1. The law of lengths states that if the tension, density and diameter of a string are held constant, the frequency is inversely proportional to the length and the change in frequency of a wave produced by the string with a change in the length of that string is given by:

$$\nu_1/\nu_2 = \ell_2/\ell_1$$

2. The law of diameters states that if the tension, density and length of a string are held constant, the frequency is inversely proportional to the diameter and the ratio of the frequencies of two strings with different diameters is given by:

$$\nu_1/\nu_2 = d_2/d_1$$

3. The law of tensions states that if the length, density and diameter of a string are held constant, the frequency is directly proportional to the square root of the tension on the string, and the change in frequency with the change in tension on the string is given by:

$$\nu_1/\nu_2 = (F_1)^{1/2}/(F_2)^{1/2}$$

4. The law of densities states that if the length, tension and diameter of a string are held constant, the frequency is inversely proportional to the square root of its

density, and the ratio of the frequencies of two strings with different densities is given by:

$$\nu_1/\nu_2 = (\rho_2)^{1/2}/(\rho_1)^{1/2}$$

"Fundamental
frequency and
harmonies"
page 300

String harmonics: Any string can be set in vibrational motion. However, to produce a standing wave, the tension in the string must be adjusted so that:

$$v = (F/\mu)^{1/2}$$

where v is the velocity of the wave in the string, F is the tension, and μ is the mass per unit of length of the string. The wavelength and frequency of the standing wave are related by $\lambda\nu = v$, so that:

$$\nu = (1/\lambda)(F/\mu)^{1/2}$$

In most musical string instruments the ends are fixed and the tension on the string is fixed. Waves reflected back from the fixed ends set up standing waves in the string. Since the amplitude of the wave is zero at the two fixed ends they act as nodes. The distance between the nodes is the length of the string, ℓ, and the wavelength of the fundamental standing wave is 2ℓ:

$$\lambda = 2\ell \qquad \text{therefore} \qquad \ell = \lambda/2$$

Overtones will be multiples of the fundamental:

$$\ell_n = n\lambda/2$$

where n is an integer that identifies the harmonic and $\lambda/2$ is the wavelength of the fundamental (first harmonic) for the string, and:

$$\nu_n = (1/n\lambda)(F/\mu)^{1/2}$$

Ex. 11-12
page 301

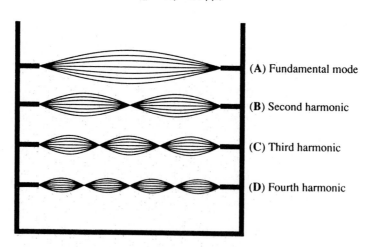

(A) Fundamental mode

(B) Second harmonic

(C) Third harmonic

(D) Fourth harmonic

Figure 4.10-4
Vibrational motion of the first four harmonics of a single string.

• Pipes: Musical instruments that use pipes include organs and horns. In string instruments the string vibrates first, the energy of the wave produced is then transferred to the molecules of the air and the compression-rarification air waves strike the ear. In pipes the column of air inside the pipe is analogous to the string. It vibrates first and then transfers its energy to the air molecules outside of the pipe. A closed end of a pipe acts like a node. The wave is reflected from its fixed boundary. In contrast, an open end of a pipe acts as an antinode, a point of maximum amplitude for the wave. Because of this difference the fundamental frequency of a closed-end pipe will be different from that of an otherwise identical open-ended pipe.

Open pipes are pipes that are open to the air at both ends. This produces an air column with free boundaries at the open ends because the air is free to expand out of the pipe. The longitudinal standing waves produced are analogous to those in strings except that the positions of all the nodes and antinodes are exactly reversed. The fundamental harmonic has antinodes at each open end and a single node in the center of the pipe; the wavelength is still twice the length of the pipe:

$$\lambda = 2\ell$$

figure 12-7
page 318

All harmonics of the fundamental tone are possible in an open-end pipe.

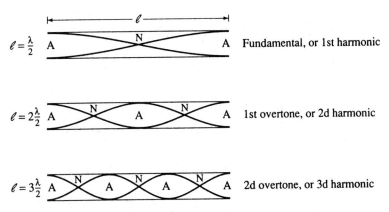

Figure 4.10-5
The first three harmonics for an open pipe. Nodes are N and antinodes are A.

Closed pipes are pipes with one closed end and one open end. The closed end provides a fixed boundary and the open end acts as a free boundary. A standing wave will always have a node at the closed end and an antinode at the open end. By definition, the distance between nodes is $\lambda/2$; the distance between a node and an adjacent antinode must be half this distance or $\lambda/4$. The fundamental wavelength must be four times the length of the closed end pipe:

$$\lambda = 4\ell$$

This is twice the λ of the fundamental in an open-end pipe. Since the wavelength and the frequency of a wave are inversely proportional, the frequency of the fundamental tone in a closed-end pipe must be half that in an open-end pipe of the same dimensions.

Not all of the possible harmonics are allowed in the closed-end pipe. Standing waves occur only for odd (noneven) values of $\lambda/4$; therefore, only odd harmonics of the fundamental tone are produced in closed-end pipes.

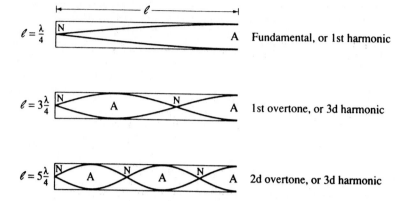

Figure 4.10-6
The first three allowed harmonics for a closed pipe. Nodes are N and antinodes are A.

Practice Problems

Ex. 12-4
page 313

"Speed of sound
in air"
page 309

1. The fog horn of a ship echoes off an iceberg in the distance. If the echo is heard 5.00 seconds after the horn is sounded, and the air temperature is −50.0°C, how far away is the iceberg?

   A. 200 m
   B. 750 m
   C. 825 m
   D. 900 m

2. What is the intensity, in W/m², of sound with a sound level of 20 dB?

   A. 10^{-12} W/m²
   B. 10^{-10} W/m²
   C. 1 W/m²
   D. 10 W/m²

3. What is the sound level of a wave with an intensity of 10^{-3} W/m²?

   A. 30 dB
   B. 60 dB
   C. 90 dB
   D. 120 dB

4. A man drops a metal probe down a deep well drilling shaft that is 3920 meters deep. If the temperature is 25°C, which is the closest estimate of how long it takes to hear the echo after the probe is dropped?

   A. 5.0 s
   B. 10 s
   C. 20 s
   D. 30 s

5. At 0°C, approximately how long does it take sound to travel 5.00 km?

   A. 15 s
   B. 30 s
   C. 45 s
   D. 60 s

6. If the speed of a transverse wave of a violin string is 12 meters per second and the frequency played is 4.0 Hz, what is the wavelength of the sound?

   A. 48 m
   B. 12 m
   C. 3.0 m
   D. 0.33 m

7. Two pulses, exactly out of phase, travel towards each other along a string as indicated below:

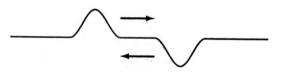

The phenomenon that occurs when the pulses meet is called

   A. refraction.
   B. reflection.
   C. diffraction.
   D. interference.

8. In which medium will the speed of sound of frequency 10^3 Hz be greatest?

able 12-1
page 309

    A. air at 25°C
    B. water at 25°C
    C. jello at 25°C
    D. iron metal at 25°C

9. If two identical sound waves interact in phase, the resulting wave will have

    A. a shorter period.
    B. a larger amplitude.
    C. a higher frequency.
    D. a greater velocity.

10. A stereo receiver has a power output of 50 W at a frequency of 1000 Hz. If the frequency is decreased to 100 Hz, the output decreases by 10 dB. What is the power output at 10 Hz?

    A. 5.0 W
    B. 10 W
    C. 50 W
    D. 100 W

Ex. 11-9
page 291

11. What is the speed of a longitudinal sound wave in a steel rod if Young's modulus for steel is 20×10^{10} N/m^2 and the density of steel is 8 10^3 kg/m^3?

    A. 4.0×10^{-8} m/s
    B. 5.0×10^3 m/s
    C. 25×10^6 m/s
    D. 2.5×10^9 m/s

"Beats"
page 323

12. If two frequencies emitted from two sources are 48 and 54 vibrations per second, how many beats per second are heard?

    A. 3
    B. 6
    C. 9
    D. 12

13. The frequency heard by a detector is higher than the frequency emitted by the source. Which of the statements below must be true?

    A. The source must be moving away from the detector.
    B. The source must be moving towards the detector.
    C. The source and the detector may be moving towards each other.
    D. The source may be moving away from the source.

Answers and Explanations

1. **B** The normal speed of sound at 0°C in air is about 330 meters/second. The speed changes by 0.6 m/s for each change of 1°C, therefore, the speed in air at −50°C is:

$$330 \text{ m/s} - 50°C(0.6 \text{ m/s°C}) = 300 \text{ m/s}$$

In 5 seconds the sound wave traveled a round trip distance of (300 m/s)(5 s) = 1500 m, therefore, the iceberg is 1500/2 = 750 meters away.

2. **B** $\beta = 10 \log I/I_0$. The threshold of hearing is 0 dB = 10^{-12} W/m^2. Since each 10 dB represents an order of magnitude of intensity, 20 dB is two orders of magnitude (100 ×) greater than the threshold intensity. This gives it an intensity of 10^{-10} W/m^2.

3. **C** $\beta = 10 \log I/I_0 = 10 \log (10^{-3}/10^{-12})$
$= 10 \log 10^9 = 10 \times 9 = 90$ dB

4. **D** There are two parts to the problem. First you must find the length of time it takes the probe to fall to the bottom of the shaft, and then you need the amount of time it takes sound to travel back that same distance:

$$y = gt^2 \text{ therefore } t = (y/g)^{1/2}$$
$$= [3920 \text{ m}/9.8 \text{ m/s}^2]^{1/2}$$
$$= (400 \text{ s}^2)^{1/2} = 20 \text{ s}$$

The probe free falls for 20 seconds before hitting the bottom of the well.

    Next, adjust the speed of sound in air for the 25°C. The velocity of sound in air increases by about 0.6 m/s °C. At 25°C, the speed of sound is 15 degrees faster, or 345 m/s. At that speed it takes about 11 seconds for the sound to return up the shaft. The total round trip time is 20 + 11 = 31 seconds. This makes **D** the closest choice. Actually, once the free fall time of 20 seconds is determined, choices A, B, and C can be eliminated because they are less than the free fall time alone.

5. **A** At 0°C the speed of sound in air is 330 m/s, therefore, it takes:

$$(5.00 \times 10^3 \text{ m})/(330 \text{ m/s}) \sim 15 \text{ s}$$

6. **C** The speed is equal to the product of the velocity and wavelength, therefore:

$$\lambda = v/\nu = (12 \text{ m/s})/4 \text{ s}^{-1} = 3 \text{ meters}$$

7. **D** By definition, two waves propagated through the same medium can be superposed to produce interference patterns.

8. **D** The frequency of the sound wave is not relevant to the question. Sound waves are fastest in solids. The stiffer the solid the faster the propogation.

9. **B** Two waves are in phase if their crests and troughs coincide. The amplitude of the resulting wave is the algebraic sum of the amplitudes of the two waves being superposed at that point. Therefore, the amplitude is doubled.

10. **A** Choices C and D can be eliminated immediately. You must have less power than at 1000 Hz in order to get the decrease in sound level. The intensity is directly proportional to the power. A 10 dB drop is a drop of one order of magnitude. Decreasing 50 W by an order of magnitude gives 5 W.

11. **B** The relationship between speed and Young's modulus is:

$$\begin{aligned}
v &= (Y/\rho)^{1/2} \\
&= [(20 \times 10^{10} \text{ kg m/s}^2)/(8 \times 10^3 \text{ kg/m}^3]^{1/2} \\
&= (25 \times 10^6 \text{ m}^2/\text{s}^2)^{1/2} \\
&= 5 \times 10^3 \text{ m/s}
\end{aligned}$$

12. **B** $\nu_{beat} = \nu_1 - \nu_2 = 54 - 48 = 6$ beats

13. **C** This is the Doppler effect. Since the frequency is shifted to a higher value the source and the detector must be getting closer together. This eliminated choices B and D. A is eliminated because it is not mandatory that the source be the device moving. The same effect is accomplished by holding the source steady and moving the detector towards it.

Key Words

audio range	frequency	reflections
beats	harmonic composition	sonic spectrum
closed pipes	intensity	sound level
decibel	laws of strings	string harmonics
diffraction	loudness	threshold of hearing
Doppler effect	open pipes	timbre
echo	pitch	

4.11 Waves: Light and Optics

A. Electromagnetic Radiation

Ch. 22
page 571

Electromagnetic radiation, EMR, has the following properties:

1. EMR can be transmitted through a vacuum with a constant speed. Unlike mechanical waves which require the presence of a physical medium, the radiant energy of heat, light and electricity can be propagated through free space (a vacuum), in the form of electromagnetic waves.

2. The energy of an electromagnetic wave is equally divided between an electric field and a mutually perpendicular magnetic field. Both fields are perpendicular to the direction of propagation; therefore, all electromagnetic waves are transverse waves.

3. All EMR waves travel through a vacuum with the same speed, $c = 3.00 \times 10^8$ m/s, called the speed of light.

4. When EMR waves travel through a physical medium, they no longer travel at the constant speed of light. The speed of a particular wave is proportional to its frequency:

$$\lambda \nu = v$$

The text uses (f) for frequency rather than the Greek letter (ν).

As the frequency decreases, the speed decreases from its maximum value of c. Any given frequency of light does not change; therefore, the change in velocity produces a change in wavelength. This is the explanation for the phenomenon of **dispersion.**

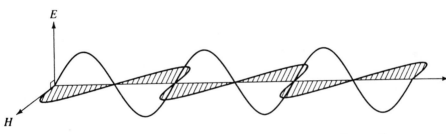

Figure 4.11-1
Direction of propagation of a wave along with its mutually perpendicular electric, *E*, and magnetic, *H*, fields.

Sec. 27-3
page 722

- **Planck's equation:** The energy in the electromagnetic spectrum can be transmitted as waves or as equivalent particles called photons. A photon is a massless particle that represents a discrete quantity of energy, called a quantum of energy.

  The amount of energy carried by the photon is related to the frequency or the wavelength of the electromagnetic radiation by Planck's equation:

$$E = h\nu = hc/\lambda$$

where $h = 6.63 \times 10^{-34}$ J s (called Planck's constant) and $c = 3.00 \times 10^8$ m/s is the speed of light in a vacuum.

Sec. 27-6
page 727

- **Matter waves:** All material bodies generate electromagnetic radiation. Because this radiation is produced by matter it is called matter waves. Because the existence of matter waves was first theorized by the physicist Louis de Broglie, these waves are also called de Broglie waves. The wavelength of the radiation produced is inversely proportional to the momentum, *mv*, of the body. The proportionality constant is Planck's constant:

$$\lambda = h/mv$$

For heavy bodies (large mass, *m*) or fast moving bodies (high speed, *v*) the momentum will be large and the wavelength of the corresponding matter wave will be very short. The matter waves produced by macroscopic bodies (such as planets, cars, people) are so short they fall outside of the spectral range that can currently be measured. Matter waves for microscopic bodies (atoms, molecules, electrons) are observable. The wavelength associated with a constant mass particle will vary with the velocity of the particle. Therefore, as the kinetic energy of a free electron increases, the wavelength of the matter wave it generates will become shorter.

Ex. 27-6
page 728

Optional Aside: Matter waves associated with macroscopic bodies exist but are not necessarily detectable. For example, the sun has a mass of approximately 2.0×10^{30} kg and rotates about the center of our galaxy with a velocity of approximately 250 km/s. The de Broglie wavelength associated with this solar motion is:

$$\lambda = h/mv = (6.63 \times 10^{-34} \text{ J s})/(2.0 \times 10^{30} \text{ kg})(250 \text{ km/s}) = 1.3 \times 10^{-69} \text{ m}$$

These matter waves generated by the sun have never been and can never be detected. Since waves are generally detected by diffraction, which occurs when they pass through slits or around obstacles whose dimensions are of the same magnitude as the wave's wavelength, the solar matter waves will always fall outside of the range we can measure because their wavelengths ($\sim 10^{-69}$ m) are smaller than the smallest constituents of matter. This means there are no slits or obstacles possible with dimensions similar to those of the waves.

- **Polarization:** All electromagnetic radiation is propagated as transverse waves which travel radially outward in all directions from their source. Because the waves are transverse, they can oscillate along any of an infinite number of planes perpendicular to the line of propagation.

An unpolarized wave is one whose direction of oscillation changes randomly with time. Unpolarized waves can be polarized.

A polarized wave oscillates in only one plane. By convention, the direction of polarization is the direction of the electric field vector. Light that oscillates only in one plane is called plane polarized light.

A polarizer is a device or substance that selects only one of the possible planes and blocks all others. Polarization can only apply to transverse waves because longitudinal waves have no oscillations perpendicular to the line of propagation.

- **Spectral regions:** Electromagnetic radiation cannot be observed directly. In order to be detected and measured, EMR must interact with matter and change some observable property of the matter. That is, in the interaction the EMR energy is transformed into some other energy form such as kinetic, potential, thermal, chemical, electrical, etc., that can be observed and measured.

The range of energies of the electromagnetic waves make up the **electromagnetic spectrum,** which is subdivided into various regions whose energies are associated with specific types of interactions with matter.

figure 22-10
page 579

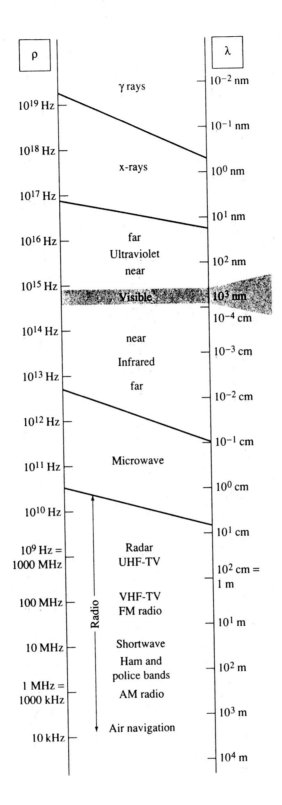

Figure 4.11-2
Electromagnetic radiation spectrum.

1. **Power wave** region: These waves are produced by electric generators and transferred along transmission lines. These are low-energy, long-wavelength, low-frequency waves.

2. **Radio wave** region: This covers a large region of the spectrum with wavelengths of a few millimeters to approximately 10 kilometers. These include the waves that carry television and radio signals. It also includes a region called the microwave region.

   The **microwave** region: This region is at the high-energy end of the radio wave region. Both radio and microwaves can penetrate the human body and produce local heating. At the molecular level, microwaves cause molecules to change their rates of rotation.

3. The **infrared** region: This region starts just above the region typically used for communications (radio and microwave regions) and extends up to the low-energy

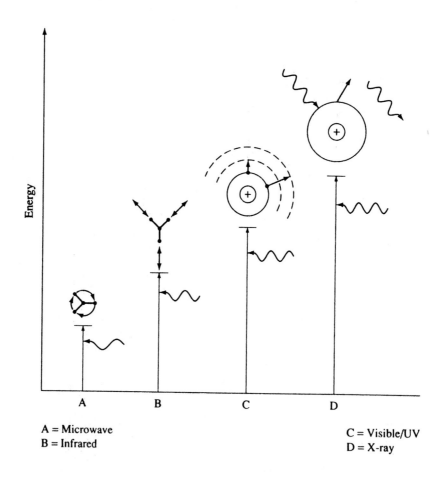

A = Microwave
B = Infrared

C = Visible/UV
D = X-ray

Figure 4.11-3
The interaction of different types of electromagnetic radiation with matter.

end of the visible region (red light). Humans perceive IR radiation as heat. At the molecular level, IR radiation produces molecular vibrations.

4. The **optical** region: This region has three major subdivisions: the near infrared, the visible and the near ultraviolet. Energies in this region cause electrons to change orbitals.

5. The **ultraviolet** region. These rays and rays of higher energy cause the ionization of atoms. This is also the region responsible for sun tans and skin cancer.

6. **X-rays:** These are produced when a metal target is hit by electrons accelerated through a high voltage drop. As the electron beam approaches the atoms in the metal lattice it is repelled by the atom's electron cloud and decelerates. Most of the kinetic energy lost by the electrons in the beam goes into increasing the thermal energy of the lattice atoms. However, approximately 1% is emitted in the form of x-rays.

   X-rays are used to expose photographic materials. Soft tissues are more transparent to x-rays than bone; therefore, bones tend to absorb the radiation and cast their shadow on the film.

7. **Gamma rays:** These are high-energy waves produced by radioactive nuclei.

   Note: The boundaries between regions of the electromagnetic spectrum are not sharp.

B. The Visual Spectrum

Ch. 23
page 589

The **visual spectrum** is the region of the optical spectrum that can be perceived by the human eye. Each component wavelength is called a color. When all the colors are present in a sample, the individual components cannot be distinguished and "white light" is produced.

Sec. 23-1
page 590

- **Light rays:** For convenience a light wave is represented by a straight line coincident with the direction of propagation of the wave.

- **Attenuation** of wave energy: When light passes from a vacuum into any material medium, or when it comes to a boundary between two media, its energy is attenuated (becomes progressively weaker) because of two effects:

  1. **Absorption:** Part of the light energy is absorbed by molecules of the physical medium it travels through and by the molecules of a boundary that it impinges upon. Generally the thermal energy of the molecules is raised by this process.

  2. **Scattering:** Part of the light energy is scattered in all directions at the boundary. A beam originally traveling in one direction is scattered along many directions. Since the total energy is constant, it is now split among a number of beams so that no scattered beam is as intense as the original beam.

Sec. 23-3
page 592 The
text does not give
this specific
formula.

The intensity of reflected light at a boundary between two media depends on the indices of refraction, n_1 and n_2, of the two media. In going from medium 1 to medium 2 the ratio of the intensity of the reflected light, I_r, to the intensity of the incident light, I_i, is given by:

$$I_r/I_i = [(n_2 - n_1)/(n_2 + n_1)]^2$$

This equation is valid for light rays that are normal to the surface (angle of incidence = 0°). Light loses about 4% of its intensity each time it is reflected from a mirror.

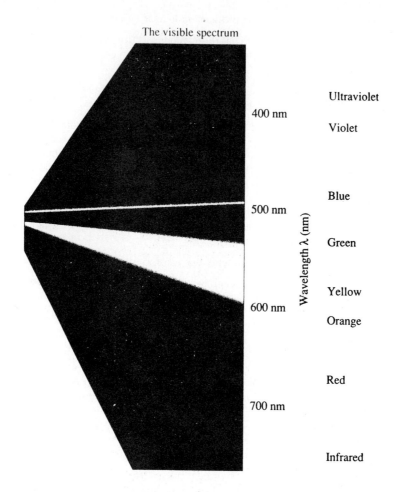

Figure 4.11-4
The visible region of the spectrum.

C. *Reflection*

Reflection occurs when light rays reach a boundary between two media and are bounced back into the original medium. The rays approaching the boundary are called the incident rays. Those bouncing off the surface are called reflected rays.

- Regular reflection occurs when each incident ray produces a single reflected ray. Scattering is negligible.

figure 23-6 page 593   **Specular surfaces** are highly polished boundaries that cause regular reflection of incident light. The image of a luminous object produced by a specular surface is sharp. A mirror is an example of a specular surface.

- The **laws of reflection** state that:

  1. The angle of incidence is equal to the angle of reflection. These angles are typically measured with respect to the normal, which is a line perpendicular to the surface at the point where the light ray strikes the surface.

2. The incident ray, the reflected ray and the normal to the reflecting surface all lie in the same plane.

Normal reflection occurs when the specular surface is a smooth plane. Normals at different points on the surface are parallel with each other. Therefore, parallel incident rays will produce parallel reflection rays.

Diffusion occurs if the surface is not a smooth plane. That is, normals to various points will not all be parallel with each other. Therefore, parallel incident rays will produce nonparallel reflection rays.

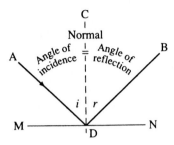

Plane surface

Figure 4.11-5

The law of reflection.

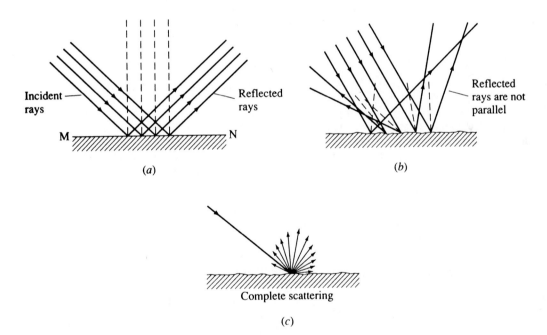

Figure 4.11-6
(a) Normal reflection (b) diffusion and (c) scattering.

- Images formed by reflection: When rays from an object are reflected, they form an image of the object. The brain interprets the position of the image by following the lines of the reflected rays back to some point where they converge or appear to converge.

A **real image** is formed by rays that actually pass through the image point. The image point is the location of the image.

"Real and virtual images" page 593

1. A real image can be projected onto a screen placed at the image point.

2. Real images are inverted with respect to the object.

3. Real images can either be magnified or reduced with respect to the size of the object.

A **virtual image** occurs when rays appear to diverge from the image point. The eye follows the diverging rays backward and extrapolates them to the point where they appear to meet (converge). The rays do not actually pass through the image point.

1. Virtual images cannot be projected on a screen.

2. Virtual images are erect with respect to the object.

3. Virtual images can be enlarged or reduced in size with respect to the size of the object.

figure 33-7 page 593

• **Plane mirrors** are flat reflecting surfaces. The image of an object in front of a plane mirror appears to be behind the mirror. Plane mirrors form virtual images because the image point is "behind" the mirror. The reflected rays never actually pass through the image point. They are reflected at the mirror surface.
    Images formed by plane mirrors:

1. are virtual

2. are erect

3. are the same size as the object

4. appear as far behind the mirror as the object is in front of the mirror

5. are reversed left to right

• **Spherical mirrors** have a curved reflecting surface.

figure 23-11 page 595 Notice that the "vertex" is not specifically identified in the text. But it is labeled on the diagrams with an (A).

The center of curvature, C, is the center of the sphere of which the mirror forms part of the surface. The radius of the sphere is called the radius of curvature, R.
    The aperture is the portion of the sphere making up the mirror.
    The vertex, V, is the geometric center of the mirror. This point lies within the body of the mirror, not on its surface.
    The principal axis is the diameter of the sphere that passes from the center of curvature through the vertex.
    A secondary axis is any other radius drawn from the center of curvature to the surface of the lens.
    A normal, N, to the surface is any radius drawn from the center of curvature to the point of incidence on the surface of the lens. The normal is perpendicular to a tangent to the surface at that point. For concave mirrors normals are radii. For convex mirrors, normals are extensions of the radii beyond the outer surface of the sphere.
    The principal focal point, F, is the point on the principal axis to which parallel incident rays converge, or appear to converge after being reflected.
    The focal length, f, is the shortest distance from the focal point to the mirror surface. The focal length of the principal focus is half the radius of curvature of the sphere.

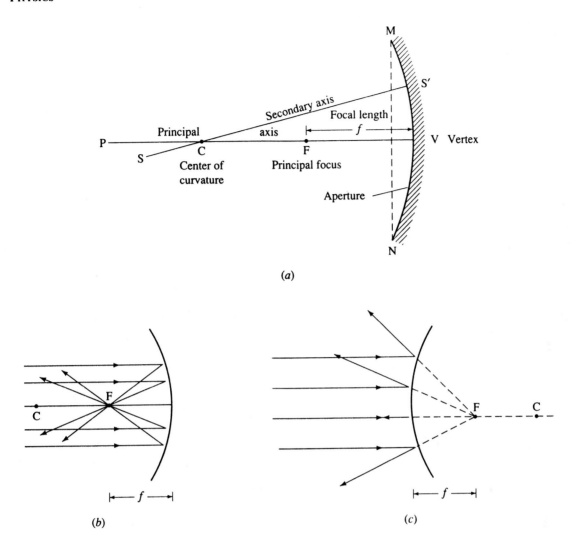

Figure 4.11-7
(a) Terms of spherical mirrors. Spherical surfaces and focal points for (b) a concave mirror and (c) a convex mirror.

A **concave mirror** has its reflecting surface on the inner surface of the sphere. Concave mirrors are converging mirrors. The principal focal point lies in front of the mirror, so it has a positive focal length:

$$f = R/2$$

The image formed depends on the distance of the object from the mirror surface. There are six general cases, as illustrated in Figure 4.11-8.

Case 1: The object is an infinite distance away from the mirror and produces parallel incident rays. If these rays are parallel to the principal axis, they will converge at the focal point. The image will be a point at the principal focal point.

Case 2: The object is at a finite distance beyond the center of curvature, C. The image is real, inverted, smaller than the object and located between the principal focal point, F, and C.

Case 3: The object is at C. The image is real, inverted, the same size as the object and located at C.

Case 4: The object is between C and F. The image is real, inverted, enlarged and located beyond C.

Case 5: The object is at F. No image is formed. If the object were a point, all the reflected rays would be parallel.

Case 6: The object is between F and the mirror. The image is virtual, erect, enlarged and located behind the mirror.

Sec. 23-4
page 594

A **convex mirror** has its reflecting surface on the outer surface of the sphere. Incident rays that are parallel to the principal axis diverge on being reflected from the mirror surface.

Convex mirrors are diverging mirrors. The principal focus lies behind the mirror, so its focal length is negative:

$$f = -R/2$$

The image produced by convex mirrors is always virtual, erect, smaller than the object, and located "behind" the mirror between the vertex and the principal focal point. Image size increases as the object moves closer to the mirror but never becomes as large as the original object.

See Eq. 23-3
page 598 for
different
notation.

The mirror equation gives the position of the image formed. The equation works for both concave and convex mirrors:

$$1/f = 1/p + 1/q$$

where f is the focal length, p is the distance of the object from the mirror, and q is the distance of the image from the mirror. Remember that f is positive for concave and negative for convex mirrors.

The **magnification, M,** for any optical system, mirror or lens, is the ratio of the size of the image, s_i, to the size of the object, s_o:

$$M = s_i/s_o = -q/p$$

where q and p are the distances respectively of the image and object from the mirror.

If M is positive, the image is erect.

If M is negative, the image is inverted.

"spherical
aberration"
page 596

Spherical aberration is a common problem with spherical mirrors and lenses with large apertures. Parallel rays incident on the outer edges of the mirror are not reflected through the principal focal point and instead converge at points between F and the mirror surface. This decreases the clarity of the image. This aberration can be reduced or eliminated by:

1. reducing the aperture size of the mirror. Spherical aberrations are negligible for mirrors with apertures of 10° or less.

Sec. 25-6
page 667

2. replacing the spherical surface with a parabolic surface.

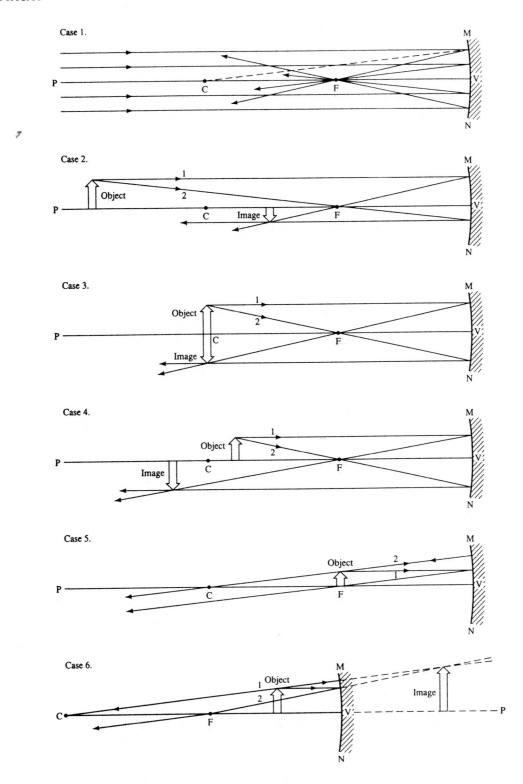

Case 1.

Case 2.

Case 3.

Case 4.

Case 5.

Case 6.

Figure 4.11-8
The six general cases for the relationship between object distance from mirror
surface and the image formed.

D. Refraction

Sec. 23-5
page 601   **Refraction** is the bending of light as it crosses obliquely from one medium into another. It arises from the difference in speed of light in different media. The slower the speed in a particular medium, the more the light will be bent upon entering that medium. The attenuation of speed in a given medium depends on the frequency of the light. The higher the frequency, the more the wave will be slowed.

Optional Aside: The speed of light in a transmitting medium is not the same as its speed in a vacuum. The attenuation of speed depends on a physical property of the medium called its optical density. The greater the optical density, the slower the speed of light in the medium. At the interface between two media of different optical densities the speed of light changes and the light ray "bends" (changes direction on passing from one medium into the next). The optical density is measured by the refractive index of the medium.

- The **refractive index,** n, of a transparent substance is the ratio of the speed of light in a vacuum, c, to its speed in the substance, v:

$$n = c/v$$

  The speed of light in air is essentially the same as in a vacuum so the refractive index is sometimes defined as the ratio of the speed of light in air to that in the substance. Since v can never exceed c, n is always greater than (or equal to) one.

- **Snell's law:** The refractive index equation above requires that you know the speed of light in the substance in order to calculate its refractive index. This information, however, is not always easily obtainable. The physicist Snell defined the refractive index in terms of the incident and refractive angles a beam made with respect to the normal to the surface:

$$n = \sin i/\sin r = v_1/v_2$$

  This is written in a more general form that allows you to compare two media:

$$\sin \theta_1 / \sin \theta_2 = v_1/v_2 = (c/n_1)/(c/n_2) = n_2/n_1$$

  or

$$n_1 \sin \theta_1 = n_2 \sin \theta_2$$

- **Laws of refraction:** Refraction can be summarized in three laws:

  1. The incident light ray, the refracted light ray and the normal to the boundary between the two media all lie in the same plane.

Ex. 23-5
page 602
  2. The index of refraction for any given medium is a constant and independent of the angle of incidence.

  3. A light ray passing obliquely between two media with different indices of refraction (optical densities) will be bent.
     —If the ray enters the medium with the larger index of refraction, the ray will be bent toward the normal because the light has been slowed to a greater extent.
     —If the ray enters the medium with the smaller index of refraction, the ray will be bent away from the normal because the light has been sped up.
     —For a transparent medium with parallel walls, such as a cube of glass, the path of a ray exiting from the glass back will be parallel to its path before entering the glass. See Figure 4.11-9.

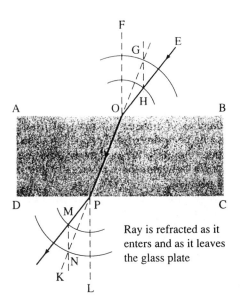

Figure 4.11-9

Internal reflection: When light reaches a boundary between two media with different optical densities, some of the light is reflected. The greater the angle of incidence, the greater the amount of light reflected at the boundary. For light in the medium with the greater refractive index there will be a critical angle, θ_c, that produces an angle of refraction of 90°. Under these conditions, the light ray skims along the interface between the two media. If the angle of incidence exceeds the critical angle, the ray will be completely reflected at the boundary. This is called internal reflection. Snell's law relates the critical angle to the refractive index:

$$\sin i_c = 1/n$$

Sec. 23-6
page 602

Ex. 23-6
page 603

See figures 23-21
and 23-22 page
604 for
application of
total internal
reflection.

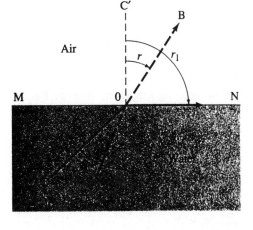 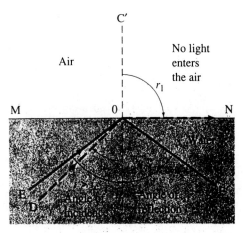

Figure 4.11-10
The critical angle and internal reflection.

• Dispersion is the separation of ordinary white light into its component colors. It is a direct consequence of refraction. In a vacuum or in air all the components travel with the same speed. When the light passes into a physical medium, light with different wavelengths will travel with different speeds and therefore get bent to different extents in a given medium. The lower the energy of the wave, that is, the longer its wavelength or lower its frequency, the less it is bent by a medium.

Prisms are solid triangles of transparent material used to disperse light. The material of the prism has a higher index of refraction than air so light waves will be bent toward the normal on entering the prism and away from the normal on exiting the prism. Because of the angle between the two faces of the prism, the entering and exiting light rays are not parallel.

High-energy violet light is refracted to a greater extent than the lower-energy red light. The light dispersed by a standard prism will form a rainbow with red light on top and violet light on the bottom.

Sec. 24-4
page 627

(*a*)

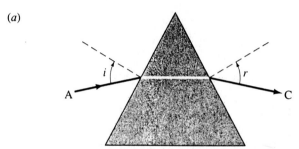

The path of a light ray through a prism

(*b*)

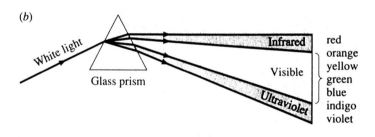

Figure 4.11-11
(*a*) Prism showing the relation between the entering and exiting rays. (*b*) The rainbow spectrum produced when white light is dispersed.

Sec. 23-7
page 605

E. Optical Lens

An optical lens is any transparent material that can be used to focus a transmitted beam of light to form an image of an object.

—Rays parallel to the principal axis will be refracted so that they converge or appear to converge at the principal focal point.

—The position of the principal focus on the axis depends on the index of refraction of the lens material. This is in contrast to spherical mirrors where the principal focus is midway between the mirror surface and the center of curvature.

—A lens can focus light because of the refractive properties of its material. Light rays go through two boundaries; one when entering the lens and the second when leaving the lens. Therefore, both surfaces of the lens must be described.

The text makes
reference to these
as "thin lenses"
because (thick)
lenses require a
different
analytical
analysis.

—Spherical lenses consist either of two curved, nonparallel sides, or one curved and one planar side. Each curved side will have a center of curvature. The principal axis passes through each center of curvature for the lens. For a symmetrical lens the radius of curvature is identical for both sides and the optical center coincides with the geometric center of the lens.

—Images formed on the same side of the lens as the object are virtual images. Images formed on the side of the lens opposite to the object are real images.

—There are two types of lenses, converging and diverging.

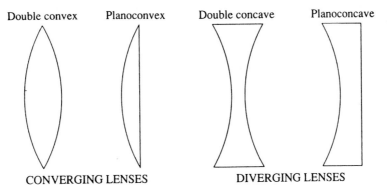

Figure 4.11-12

- **Converging lenses** are thicker in the middle than at the edges. Double convex and plano-convex are examples of converging lenses. Parallel rays converge at the principal focal point. Because rays actually go through the focus, that focus is called a real focus. Real images form at a real focus. The real focus is on the opposite side of the lens from the object.

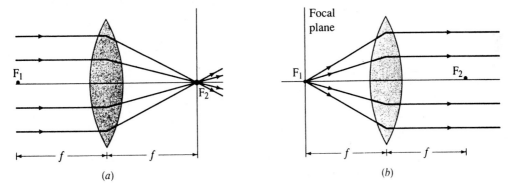

Figure 4.11-13

A double convex converging lens.

For converging lenses the image produced depends on the location of the object. There are six cases, as illustrated in Figure 4.11-14:

Case 1: The object is at infinity; the image formed is a point coincident with the real focal point.

Case 2: The object is more than two focal lengths away; the image is real, inverted, reduced in size and located between F and 2F on the opposite side of the lens.

Case 3: The object is exactly two focal lengths away at 2F; the image is real, inverted, the same size as the object and located at 2F on the opposite side of the lens.

Case 1.

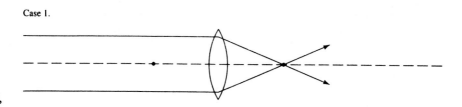

Figures 23-28, 23-29 and 23-30 in the text are diagrams showing how ray tracing can be used to locate the position, size and orientation of the images.

Case 2.

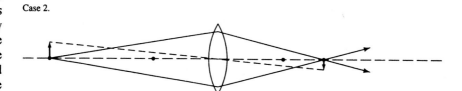

Case 3.

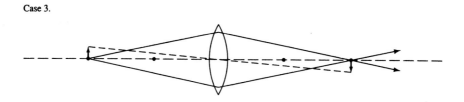

Case 4.

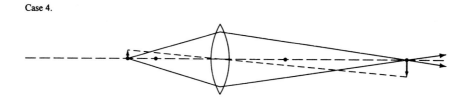

Case 5.

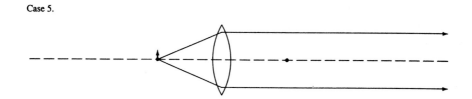

Case 6.

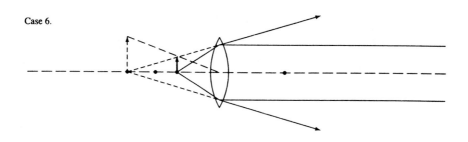

Figure 4.11-14
The six cases for image formation with a converging lens.

Case 4: The object is between the principal and second focal point, (between F and 2F); the image is real, inverted, enlarged and located beyond 2F on the opposite side of the lens.

Case 5: The object is at the principal focus, F; no image is formed because the refracted waves are parallel to each other.

Case 6: The object is between F and the lens surface; the image is virtual, erect, enlarged and located on the same side of the lens as the object.

- **Diverging lenses** are thicker at the edges than in the middle. Double concave and planoconcave are examples. Rays parallel to the principal axis are refracted and diverge on the other side of the lens. The eye follows the diverging rays and extrapolates to the point where they appear to converge. This is the principal focal point. Since no rays actually pass through this point it is called a virtual focus. Virtual images are formed on the same side as the virtual focus.

Images formed by diverging lenses are always virtual, erect and reduced in size.

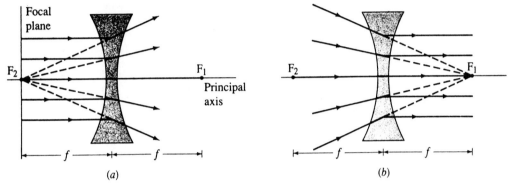

Figure 4.11-15
Images formed by diverging lenses.

Sec. 23-8
page 607

- The **thin lens equation** is equivalent to the mirror equation and gives the position of the image formed with respect to the focal length and the distance of the object from the lens. The equation works for both concave and convex lenses:

$$1/f = 1/p + 1/q$$

where f is the focal length, p is the distance of the object from the lens, and q is the distance of the image from the lens.
    —Remember that f is positive for converging lenses and negative for diverging lenses.
    —If the sign of q is positive, the image formed is real. If the sign is negative, the image is virtual.

- The magnification, M, for any optical system, mirror or lens, is the ratio of the size of the image, s_i, to the size of the object, s_o:

$$M = s_i/s_o = -q/p$$

where q and p are the distances respectively of the image and object from the mirror.
    If M is positive, the image is erect.
    If M is negative, the image is inverted.

- **Compound optical systems:** Few optical instruments contain only a single lens. Most are made of a series of lenses. However, each lens in the system behaves independently, acting as if it were the only lens present. The effect of a given lens on light determines the character of the light rays (parallel, converging, diverging, etc.) that enter the next lens in the series.

Practice Problems

1. Glass has an index of refraction of 1.50. What is the frequency of light that has a wavelength of 500 nm in glass?

   A. 1.00 Hz
   B. 2.25 Hz
   C. 4.00×10^{14} Hz
   D. 9.00×10^{16} Hz

2. Water has an index of refraction of 1.33. If a plane mirror is submerged in water, what is the angle of reflection if light strikes the mirror at an angle of 30°?

   A. less than 30° because the index of refraction for water is 1.33
   B. exactly 30°
   C. more than 30° because the index of refraction for water is 1.33
   D. no light is reflected because 30° is the critical angle for water

3. The index of refraction for water is 1.33 and that for glass is 1.50. A light ray strikes the water/glass boundary with an incident angle of 30.0° on the water side. Which statement is most likely true for its refracted angle in the glass?

   A. The angle of refraction is 26.3°.
   B. The angle of refraction is 34.7°.
   C. The angle of refraction is 30.0°.
   D. The angle of refraction is 60.0°.

4. Light is incident on a prism at an angle of 90° relative to its surface. The index of refraction of the prism material is 1.50. Which of the following statements is most accurate?

   A. The angle of refraction will be greater than 0° but less than 45°.
   B. The angle of refraction will be greater than 45° but less than 90°.
   C. The angle of refraction will be 0°.
   D. The angle of refraction cannot be determined from the information given.

The color undergoing the greater amount of refraction will have the smaller "angle of refraction." See how the "angle of refraction" is defined in Sec. 23-5 on page 601 of the text.

5. White light incident on an air/glass interface is split into a spectrum within the glass. Which statement is most accurate?

   A. Red light has the greatest angle of refraction.

   B. Violet light has the greatest angle of refraction.
   C. Yellow light has the greatest angle of refraction.
   D. The color with the greatest angle of refraction cannot be determined from the information given.

6. A real object is placed 10 cm from a converging lens that has a focal length of 6 cm. Which statement is most accurate?

   A. The image is real, erect and enlarged.
   B. The image is real, inverted and enlarged.
   C. The image is real, erect and reduced.
   D. The image is real, inverted and reduced.

Ex. 23- page 60

7. What is the focal length of a lens that forms a virtual image 30 cm from the lens when a real object is placed 15 cm from the lens?

   A. 10 cm
   B. 15 cm
   C. 30 cm
   D. 45 cm

Ex. 23- page 61

8. What is the magnification of a lens that forms an image 20 cm to its right when a real object is placed 10 cm to its left?

   A. 0.50
   B. 1.0
   C. 1.5
   D. −2.0

9. The human eye can respond to light with a total energy of as little as 10^{-18} J. If red light has a wavelength of 600 nm, what is the minimum number of red light photons that the eye can perceive?

   A. 1
   B. 2
   C. 3
   D. 5

10. Which phenomenon occurs for transverse waves but not for longitudinal waves?

    A. reflection
    B. refraction
    C. diffraction
    D. polarization

Answers and Explanations

1. **C** The velocity of light in glass can be found from the definition of refractive index, $n = c/v$. Wavelength and frequency are related to velocity by the general wave relation, $v = \lambda \nu$, therefore:

$$\nu = v/\lambda = c/n\lambda$$
$$= (3.00 \times 10^8 \text{ m/s})/(1.50)(500 \times 10^{-9} \text{ m})$$
$$= 4.00 \times 10^{14} \text{ Hz}$$

2. **B** The law of reflection is independent of the medium involved. The angle of reflection is equal to the angle of incidence for normal reflection.

3. **A** Snell's law is:

$$n_1 \sin \theta_1 = n_2 \sin \theta_2$$

$$1.33 \sin 30° = 1.50 \sin \theta_2$$

Since $n_1 < n_2$, then $\sin \theta_1 > \sin \theta_2$. For $\sin 30° > \sin \theta_2$, 30° must be greater than θ_2, for which A is the only reasonable choice.

4. **C** Snell's law is $n_1 \sin \theta_1 = n_2 \sin \theta_2$. Since the incident rays are normal to the prism surface, $\theta_1 = 0°$, and $\sin 0° = 0$. Since

$$0 = n_2 \sin \theta_2$$

$$\theta_2 = 0$$

5. **B** When light travels from one medium into another with a different index of refraction, the frequency of the light does not change.

The velocity change produces a change in wavelength. The higher the frequency, the greater the energy of the light wave and the greater the extent it is refracted by. In the visible spectrum, violet is the highest frequency light.

6. **B** The object is between one and two focal lengths from the converging lens; therefore, the image produced must be real, inverted and enlarged.

7. **C** The distance is given by the thin lens equation, $1/f = 1/p + 1/q$. Since the image formed is virtual the sign of q is negative. Therefore:

$$1/f = 1/p - 1/q = 1/15 - 1/30 = 1/30$$

$$f = 30 \text{ cm}$$

8. **D** Magnification is given by $M = -q/p = -20 \text{ cm}/10 \text{ cm} = -2.0$. The sign of q is positive because the image is real. The negative value of M means the image is inverted.

9. **C** $E = hc/\lambda = (6.63 \times 10^{-34} \text{ J s})(3.00 \times 10^8 \text{ m/s})/600 \times 10^{-9} \text{ m}) = 3.31 \times 10^{-19} \text{ J}$. This is the energy of each red photon. The number of such photons needed to produce a total of 10^{-18} J of energy is:

$$(10^{-18} \text{ J})/(3.31 \times 10^{-19} \text{ J/photon}) \sim 3 \text{ photons}$$

10. **D** Polarization can only occur with tranverse waves because the motion must be perpendicular to the direction of propagation.

Key Words

absorption
attenuation
compound optical systems
concave mirror
converging lenses
convex mirror
diffusion
dispersion
diverging lenses
electromagnetic radiation spectrum
gamma rays
infrared radiation
internal reflection
laws of reflection

laws of refraction
light rays
magnification
matter waves
microwave
normal reflection
optical radiation
Planck's equation
plane mirror
polarization
power wave
prism
radio wave
real image

refraction
refractive index
reflection
scattering
Snell's law
spectral region
specular surfaces
spherical aberrations
spherical mirrors
thin lense equation
ultraviolet radiation
virtual image
visual spectrum
X-rays

4.12 Electrostatics

Ch. 16
page 416

A. Electric Charge

Electric charge, e, is a scalar quantity with a magnitude:

$$e = 1.6 \times 10^{-19} \text{ C} = 1 \text{ esu}$$

The SI unit of electrical charge is the coulomb, C, defined as the total charge on a specific number of electrons, e^-:

$$1 \text{ C} = 6.25 \times 10^{18} \text{ } e^-$$

—Do not confuse the symbol for electric charge, e, with the very similar symbol for the subatomic particle the electron, e^- or $_1^0 e$.

—The whole charge on a body is given by the symbol q. It is an integer multiple, n, of the unit electric charge:

$$q = ne$$

—Electric charge is either negative or positive.

- Carriers of charge: The **proton,** p, carries the fundamental unit of positive charge, $+e$. The **electron,** e^-, carries the fundamental unit of negative charge. All other electric charges are integral multiples of e.

Sec. 16-2
page 418

Ions: In the atom, protons are found in the nucleus and the electrons circle the nucleus. The attractive force between the protons and the electrons decreases with increasing distance between them. The outermost or valence electrons are loosely held and are easily removed to produce ions and free electron clouds.

The number of protons in a nucleus is fixed and cannot be changed. The only way to change the net charge on an atom (or on any body) is to change the number of electrons present.

—If electrons are removed, there are more protons in the nucleus than surrounding electrons. The ion produced has a net positive charge and is called a cation.

—If electrons are added, there are more surrounding electrons than there are protons in the nucleus. The ion produced has a net negative charge and is called an anion.

In solids the relative positions of the nuclei are fixed in what is called a lattice structure. The electrons may become detached from their original nuclei and move within the lattice of cations as a cloud of electrons.

Fluids are liquids and gases. In fluids the loosely held valence electrons are transferred between nuclei to form ions. Electric charge is due primarily to the production and the motion of the resultant ions. To a lesser extent, charge can be due to a free electron cloud.

Sec. 16-3
page 419

Conductors are substances through which electric charge is readily transferred.

Most metals are good conductors because they have loosely held valence electrons which are easily removed and free to move through the lattice of the solid's nuclei.

Salts are conductors if they dissociate into their ionic components in solution. In order to dissociate, the salt must dissolve in the solvent.

Insulators are substances through which electric charge is not readily transferred.

— Most nonmetals are insulators because they hold their valence electrons too tightly to form free electron clouds.

— Molecular compounds are insulators because they do not have ionic components and remain neutral when they dissolve.

— Salts that do not dissolve cannot dissociate into their ionic components.

Electrolytes are compounds that can conduct electricity. They are ionic compounds that dissociate into their ionic components in solution. It is the ions formed that conduct charge. **Nonelectrolytes** are compounds that cannot conduct electricity. These are molecular (or nonionic) compounds or ionic compounds that do not dissolve in the given solvent. In either case no ions are formed, so no electricity can be conducted.

- The net charge, q, on a body is the difference between the total positive charge (total number of protons) and the total negative charge (total number of electrons).

Sec. 16-4
page 419

- **Electrification** is any process that produces a net electric charge on a body. It is usually produced by friction between two surfaces that causes the transfer of electrons from one surface to the other.

    — If a hard rubber rod is rubbed with fur, the rod acquires a net negative charge and the fur a net positive charge. These charges are related to the excess or deficiency of electrons produced by the friction. Electrons are mechanically transferred from the fur to the rubber. The rubber now has an excess number of electrons.

    — If a glass rod is rubbed with silk, the rod acquires a net positive charge and the silk a net negative charge.

Static electricity occurs when the electric charge is stationary; confined to a body and not moving.

Electricity is electric charge that is in motion and is not stationary.

- **Point charge:** If the net charge on a body behaves as if it were concentrated at one point on the body then it is said to act like a point charge.

B. The Law of Conservation of Charge

This law states that net charge cannot be created or destroyed. For every unit of positive charge produced (or consumed) in an interaction, a unit of negative charge must also be produced (or consumed) so that the algebraic sum of the electric charges in any closed system remains constant.

    — For reactions that convert matter to energy (or energy into matter), equal amounts of positive and negative charge are produced (or consumed).

    — For reactions that redistribute or separate charge, the algebraic sum of the electric charges must remain constant. The sum of the charges on the product side of the reaction must be the same as the sum of the charges on the reactant side.

C. Electrostatics

Electrostatics is the study of stationary electric charges. The principles of electrostatics are empirical (based on experiment rather than theory). They can be summarized by eight basic observations:

1. There are only two kinds of charges, positive and negative.

2. Opposite charges exert an attractive force on each other.

3. Like charges exert a repulsive force on each other.

4. A charged body can attract (or repel) an uncharged body.

5. Electric forces act over distance and do not require physical contact between the interacting bodies. The strength of the interacting force decreases with increasing distance between the bodies. This observation leads to Coulomb's Law.

6. When charges are separated by contact, such as rubbing two bodies together (rubber and fur), equal amounts of positive and negative charge are produced. This observation is the **Law of Conservation of Charge.**

7. The charge on a uniform solid or hollow conducting sphere is uniformly distributed over the entire surface of the body.

figure 16-25
page 430

—There is no net charge in the interior of the body. All charge is confined to the surface of the body.
—Although the charge is actually on the surface, the sphere behaves as if all the charge is concentrated at its geometric center.
—The sphere exerts no force on a charged particle placed at any point inside the sphere.

8. If the body is not a uniform sphere, the charge is still confined to its surface but the distribution varies with the curvature of the surface. Charge tends to concentrate at sharp points on the body. For example, if the body is shaped like an egg the charge will be more concentrated at the smaller, more narrow end than at the large end. (The electric field intensity is larger at sharp points.) If the concentration of charge becomes great enough, it can spontaneously discharge, producing a discharge spark as it ionizes the gas of the air around the conductor.

Sec. 16-5
page 420

• **Coulomb's law** of electrostatics states that the force of interaction, F, between two point charges, q_1 and q_2, that are at rest is directly proportional to the product of their charges and inversely proportional to the square of the distance, r^2, between them.

$$F = kq_1q_2/r^2$$

Ex. 16-1 gives an
idea of how large
such a force can
be.

In the SI system the proportionality constant k has the value:

$$k = 9 \times 10^9 \text{ N m}^2 /\text{C}^2$$

—If the force of interaction is attractive, F_{att}, the charges are opposite in sign and the sign of F_{att} is negative.
—If the force of interaction is repulsive, F_{rep}, the charges are of the same sign and the sign of F_{rep} is positive.

Optional Aside: The numerical value of k in a vacuum is 8.987×10^9 N m^2/C^2 and in air it is 8.93×10^9 N m^2/C^2. Both are approximated as 9×10^9 N m^2/C^2.

Se. 16-7
page 425

• An **electric field** is produced by a point charge and is the region of space around that charge which can exert a electric force on any other point charge that enters the region (or field).

The symbol "E"
is commonly
used for electric
field intensity as
shown in the
text.

The **electric field intensity, $\mathscr{E}$,** at any point in the field is the vector quantity that measures the force exerted on a positive test charge placed at that point:

$$\mathscr{E} = F/q$$

The SI units of $\mathscr{E}$ are the newton per coulomb, N/C.
Since F is a vector, $\mathscr{E}$ is also a vector.

—For a positive test charge, the force and electric field vectors are parallel.
—For a negative test charge the force and electric field vectors are antiparallel.

figure 16-23
page 429

Electric lines of force are used to represent the electric field surrounding a point charge or interacting between two or more point charges. The lines have the same direction as the field. By convention, the lines originate at and point away from a positive charge; they terminate at and point toward a negative charge. Force lines are drawn so that a tangent to the line gives the direction of the electric field at that point. The intensity of the electric field is proportional to the density of the lines; that is, the number of force lines per unit area normal to the field.

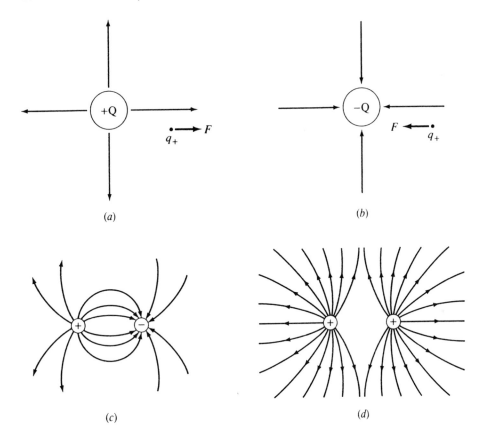

Figure 4.12-1
Direction of force, F, and field lines associated with (a) a positive and (b) a negative point charge. (c) Field lines between two oppositely charged bodies. (d) Field lines between two like charged bodies.

Sec. 17-1
page 440

• The **potential difference,** V, between two points in an electric field is the work required per unit charge to move a charge between the two points in the field:

$$V = W/q$$

The SI unit of potential difference is the volt, v, which is the ratio of one joule per coulomb:

$$1 \text{ v} = 1 \text{ J/C} = \text{N m/C}$$

1 volt is the potential difference between two points in an electric field if 1 joule of work is done in moving a charge of 1 coulomb between the two points.

—If work is done in moving a charge between two points, the two points are at different electric field potentials. The magnitude of the work done in moving the charge is the difference in the potential between the two points.

—If no work is done in moving a charge between two points, they must be at the same potential difference. They are equipotential. Since work is $W = Fd$, the potential can be expressed as:

Ex. 17-1
page 442

$$V = W/q = Fd/q$$

—The electric field, $\mathscr{E}$, can also be expressed in terms of the potential difference:

Ex. 17-2
page 444

$$\mathscr{E} = F/q = (qV/d)/q = V/d$$

The electric field intensity between two points is the ratio of potential difference between the points to their distance.

The term potential gradient refers to the change on potential difference per unit of distance.

First paragraph
on page 442
describes the
meaning of a
potential at a
point.

Grounding: Potential difference at a point is a relative term and has meaning only with respect to the assigned value of some reference point. The earth is considered to be a conductor with an electric potential arbitrarily assigned the value of zero. Since the earth is so large, it acts as a limitless source of electrons; electrons can be added to or removed from the earth without changing the planet's potential. If a conductor is connected to the earth, electrons will move between the conductor and the earth until the potential between them is equal. In other words, any conductor electrically connected to the earth must also be zero potential. Such a conductor is said to be grounded.

An **equipotential line** is one where all points on the line are at the same potential energy.

Sec. 17-3
page 444

An **equipotential surface** is one where all points on the surface are at the same potential energy. The potential difference between any two points on the surface is zero; therefore, no work is done in moving a test charge between two points on the surface. Equipotential surfaces are perpendicular to the lines of force.

- **Electric dipole:** If equal quantities of opposite charges are separated on the same body, they form an electric dipole. The centers of charge are called partial charges, q. The strength of the electric dipole is measured by the dipole moment, μ, which is the product of the partial charge q at either center, and the distance between the centers. The SI unit of the dipole moment is the Debye, D:

$$\mu = qd$$

The text uses the
symbol "p" for
the dipole
moment. Sec.
17-6 page 447.

Optional Aside: Common examples of electric dipoles are individual polar covalent bonds and the polar molecules that can result.

In covalent bonds, the bonding electrons are shared by the two atoms of the bond. The tendancy of an atom to attract these bonding electrons is measured by its electronegativity. If the two atoms of the bond have different electronegativities, the electrons will spend more time, on average, closer to the more electronegative atom, producing a polar covalent bond. Polar molecules such as H_2O have permanent electric dipoles because of the nonuniform distribution of the electrons in the covalent bonds.

The electric dipole moment is a vector quantity. The relative electronegativities of the two atoms in the bond determine the strength of the bond dipole. The geometry of the molecule determines the relative directions of the bond dipoles. If the vector sum of the bond dipoles is nonzero, the resulting molecule is polar.

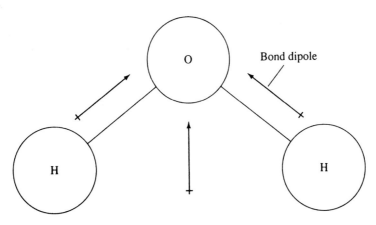

Direction of net molecular dipole

(a)

See Table 17-1
on page 448 for
typical values.

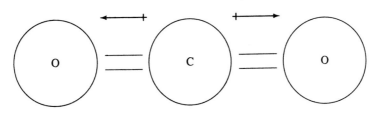

Net molecular dipole = 0 Debye

(b)

Figure 4.12-2

(a) Water is polar because the H—O bonds are polar. The two bonds are equal in magnitude but are not antiparallel, so their vector sum is a nonzero value. (b) Carbon dioxide is a nonpolar molecule. The C=O bonds are polar but antiparallel, so their effects are canceled.

Practice Problems

1. What is the potential difference between point A and point B if 10 J of work is required to move a charge of 4.0 C from one point to the other?

   A. 0.4 V
   B. 2.5 V
   C. 14 V
   D. 40 V

2. How much work is required to move an electron between two terminals whose potential difference is 2.0×10^6 volts?

   A. 3.2×10^{-13} J
   B. 8.0×10^{-26} J
   C. 1.25×10^{25} J
   D. 8.0×10^{-13} J

Ex. 17-1
page 442

3. Two electrically neutral materials are rubbed together. One acquires a net positive charge. The other must

    A. have lost electrons.
    B. have gained electrons.
    C. have lost protons.
    D. have gained protons.

4. Which diagram best represents the electric field lines around two oppositely charged particles?

    A.

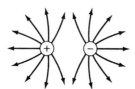

    B.

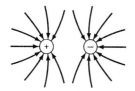

    C.

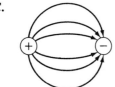

    D.

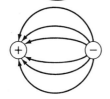

5. What is the charge on a body that has an excess of 20 electrons?

    A. 3.2×10^{-18} C
    B. 1.6×10^{-18} C
    C. 3.2×10^{-19} C
    D. 2.4×10^{-19} C

6. Two point charges, A and B, with values of 2.0×10^{-4} C and -4.0×10^{-4} C respectively are separated by a distance of 6.0 meters. What is the magnitude of the electrostatic force exerted on point A?

    A. 2.2×10^{-9} N
    B. 1.3 N
    C. 20 N
    D. 36 N

7. Two point charges A and B are separated by 10 meters. If the distance between them is reduced to 5.0 meters the force exerted on each

    A. decreases to half its original value.
    B. increases to twice its original value.
    C. decreases to one quarter of its original value.
    D. increases to four times its original value.

8. Sphere A with a net charge of $+3.0 \times 10^{-3}$ C is touched to a second sphere B, which has a net charge of -9.0×10^{-3} C. The two spheres are then separated. The net charge on sphere A is now:

    A. $+3.0 \times 10^{-3}$ C
    B. -3.0×10^{-3} C
    C. -6.0×10^{-3} C
    D. -9.0×10^{-3} C

9. Which electric charge is possible?

    A. 6.02×10^{23} C
    B. 3.2×10^{-19} C
    C. 2.4×10^{-19} C
    D. 8.0×10^{-20} C

10. If the charge on a particle in an electric field is reduced to half its original value, the force exerted on the particle by the field is

    A. doubled.
    B. halved.
    C. quadrupled.
    D. unchanged.

11. Points A, B and C are at various distances from a given point charge. Which statement is most accurate? The electric field strength is

    A. greatest at point A.
    B. greatest at point B.
    C. greatest at point C.
    D. the same at all three points.

Ex. 16-5
page 427

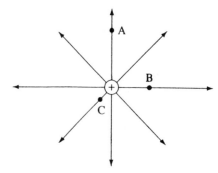

Problems 12, 13 and 14, require the use of Coulomb's Law.

12. If the magnitude of the charge on two identical charged bodies is doubled, the electrostatic force between the bodies will be

A. doubled.
B. halved.
C. quadrupled.
D. unchanged.

13. The electrostatic force between two point charges is F. If the charge of one point charge is doubled and that of the other charge is quadrupled, the force becomes which of the following?

A. $1/2F$
B. $2F$
C. $4F$
D. $8F$

14. The graph that best represents the relation between the electrostatic force F associated with two point charges and the distance d separating the point charges is

A.

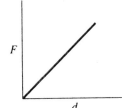

B.

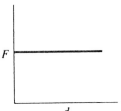

C.

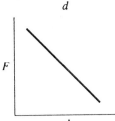

D.

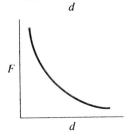

15. What is the amount of work done on an electron by a potential difference of 100 volts?

A. 6.25×10^{20} eV
B. 100 eV
C. 1.60×10^{-17} eV
D. 1.60×10^{-19} eV

16. Which of the following is a vector quantity?

A. Electric charge
B. Electric energy
C. Electric power
D. Electric field intensity

Answers and Explanations

1. **B** Potential difference between two points in an electric field is the work per unit charge required to move a charge between the two points:

$$V = W/q = \text{joules/coulomb} = \text{volts}$$
$$= 10 \text{ J/4.0 C} = 2.5 \text{ V}$$

2. **A** Work is the product of the magnitude of the charge and the potential difference it is moved through:

$$W = qV = (2.0 \times 10^6 \text{ V})(1.6 \times 10^{-19} \text{ C})$$
$$= 3.2 \times 10^{-13} \text{ J}$$

3. **B** Protons are fixed in the nucleus and cannot be transferred by friction. Electrons can be transferred by friction. Therefore, net charges are due to the transfer of electrons between two bodies. Conservation of charge means that if there is a net positive charge, one body must have lost electrons and the other body must have gained the electrons.

4. **C** Lines of electric force, by convention, originate on a positive charge center and terminate on a negative charge center.

5. **A** The elementary charge e is 1.6×10^{-19} C. Therefore, 20 electrons have a total magnitude of:

$$20(1.610^{-19} \text{ C}) = 3.2 \times 10^{-18} \text{ C}$$

6. **C** Coulomb's law:

$$F = k(q_A q_B)/r^2$$

$$F = (9.00 \times 10^9 \text{ N m}^2/\text{C}^2)(2.0 \times 10^{-4} \text{ C})(-4.0 \times 10^{-4} \text{ C})/(6.0 \text{ m})^2$$

$$F = 720 \text{ N m}^2/36 \text{ m}^2 = 20 \text{ N}$$

7. **D** From Coulomb's law, $F = k(q_A q_B)/r^2$, force is inversely proportional to the square of the distance separating the points. Decreasing the distance to half its original value means the force quadruples.

8. **B** This is an application of the law of conservation of charge. The initial net charge is:

$$(+3.0 \times 10^{-3} \text{ C}) + (-9.0 \times 10^{-3} \text{ C})$$
$$= -6.0 \times 10^{-3} \text{ C}$$

The same net charge must exist after contact. The -6.0×10^{-3} C is evenly distributed between the two spheres.

9. **B** The fundamental charge is 1.6×10^{-19} C. All net charges must be integer multiples of this value.

10. **B** The electric field strength is the ratio of the force exerted on a unit charge in the field:

$$\mathscr{E} = F/q$$

Therefore, F and q are directly proportional and linearly related.

11. **C** From Coulomb's law, force varies inversely with the square of the distance from the charge. The strength of the electric field at a point is the ratio of this force to the charge:

$$\mathscr{E} = F/q = (k(q_1 q_2)/r^2)/q$$

Therefore, $\mathscr{E}$ and r^2 are inversely proportional. The smaller the value of r, the smaller the value of r^2 and the greater the value of $\mathscr{E}$.

12. **C** From Coulomb's law, $F = k(q_q q_2)/r^2$, force is directly proportional to the product of the two charges q_1 and q_2. If both charges are doubled, their product is quadrupled and so is the electrostatic force.

13. **D** From Coulomb's law, $F = k(q_1 q_2)/r^2$, force is directly proportional to the product of the charges and since:

$$q_1 q_2 = (2q_1)(4q_2) = 8q_1 q_2$$

The product increases eight-fold and so does the force.

14. **D** Force varies inversely with the square of the distance, therefore, expect a nonlinear (quadratic) decrease in F with increase in distance, d.

15. **B** The potential difference between two points is the ratio of the work done in moving the unit charge between the two points. Therefore, $W = qV$.

16. **D** A vector has magnitude and direction. The choices A, B and C have magnitude only. The direction of the electric field vector is the same as the direction of the force.

Key Words

conductors	electrolytes	law of conservation of charge
Coulomb's law	electron	nonelectrolytes
electric charge	electrostatics	point charge
electric dipole	equipotential line	potential difference
electric fields	equipotential surface	proton
electric field intensity	grounding	static electricity
electric lines of force	insulators	
electrification	ions	

4.13 Electronic Circuits

A. Current

Sec. 18-2
page 462

Current, I, is the rate of flow of charge through a cross-sectional area of a conductor. This flow can consist of either positive or of negative charges. By convention, the direction of current flow is the direction in which the positive charges would move.

$$\text{Current} = \text{Charge/Time} \qquad I = Q/t$$

The SI unit of current is the ampere, A. One ampere equals a flow rate of one coulomb, C, per second, s.

$$1 \text{ A} = 1 \text{ C/s}$$

Three conditions are required to produce an electric current:

1. A potential difference across part of the conductor. The potential difference or voltage or emf can be produced by a generator or by a battery. It provides the acceleration to the charges.

2. A closed path or circuit through which the charges can move. If the path is open, the flow of charge is stopped and a static charge will develop instead.

3. Carriers of charge that are free to move throughout the circuit.

Solid conductors: The charge carriers are free electrons. Most metals are solid conductors.
Electrolytes in solution: The carriers are the cations (positive ions) and anions (negative ions) produced by the dissociation of the electrolyte in the solvent.
Ionized gases: Gases usually are poor conductors unless they are ionized. The carriers are ions and free electrons.

Sec. 18-8
page 471

• **Direct current,** dc, is produced if the direction in which the current flows is constant. Direct current is described by a sine wave where all points are of the same sign.

Figure 18-17
shows the
difference
between direct
current and
alternating
current.

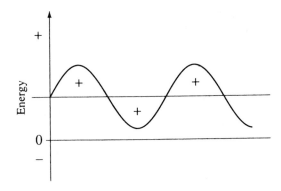

Figure 4.13-1
Sine wave representation of direct current.

• **Alternating current,** ac, is produced if the direction in which the current flows reverses direction periodically. Alternating current is described by a sine wave that changes sign every half cycle.

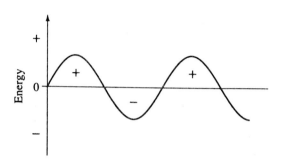

Figure 4.13-2
Sine wave representation of alternating current.

B. Resistance

Sec. 18-3
page 463

Resistance, R, is the opposition to the flow of charge provided by the material the current is flowing through.

The SI unit of resistance is the ohm, Ω, where 1 Ω = 1 volt/ampere.

• The **drift velocity,** v_D, is the average velocity of the charge carriers. It is constant for a given conductor carrying a given magnitude of current.

Optional Aside: Free electrons in a vacuum are accelerated by an electric field. In a conductor, however, the motion of the electrons is interrupted by frequent collisions with the fixed nuclei of the solid conductor. During the collisions kinetic energy is transferred from the electrons to the nuclei, increasing their vibrational motion and therefore their temperature. Because the kinetic energy is transformed irreversibly into thermal energy, collisions are a dissipative process involving a nonconservative force.

• **Laws of resistance:**

1. Resistance increases with increasing temperature for pure metals and most metallic alloys. Resistance decreases with increasing temperature for carbon, semiconductor materials and most electrolytic solutions.

2. The resistance of a uniform conductor is directly proportional to the length, ℓ, of the conductor:

$$R/\ell = constant$$

3. The resistance of a uniform conductor is inversely proportional to its cross-sectional area, A:

$$RA = constant$$

4. The resistance of a given conductor depends on the material it is composed of.

Sec. 18-4
page 465

Resistivity, ρ, summarizes the laws of resistance. The resistance of a uniform conductor is directly proportional to its length and inversely proportional to its area. The proportionality constant is the resistivity and depends only on the material and temperature of the conductor:

$$R = \rho(\ell/A)$$

The SI unit of resistivity is Ω m.

C. *Voltage*

Voltage, V, is the potential difference across a conductor. It supplies the force that accelerates the charges so they flow.

The transformation of available energy into electrical energy is accomplished by either:

—a dynamic converter, which is a rotating machine such as a generator

—a direct converter, which has no mechanical moving parts, such as a cell or battery

Sec. 18-1
page 460

- Electrochemical cells: Certain spontaneous chemical reactions, called redox reactions, involve the transfer of electrons from one reactant to another. If the two reactants are physically separated, the electrons can be transferred between them via an external circuit. The path of the current is from the reactant being oxidized, through the circuit, to the reactant being reduced. Current will flow until one of the reactants is used up. The cell must be replaced or be recharged by having the reaction reversed.

Electrodes: Generally, reactants are electrolytes and their connections to the external circuits are called electrodes. There are two types of electrodes:

1. A cathode is the electron-rich electrode; it has a net negative charge and acts as a source of electrons, supplying them to the circuit.

2. An anode is the electron-deficient electrode; it carries a net positive charge and attracts electrons, receiving them from the circuit.

The flow of charge in a cell is unidirectional, so cells are a source of direct current. Electrons flow from the cathode, through the circuit and into the anode. The circuit symbol for a single cell is:

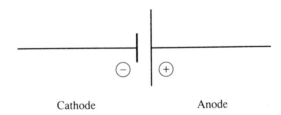

Cathode   Anode

Optional Aside: The <u>cat</u>hode attracts the <u>cat</u>ions of the electrolyte. The <u>an</u>ode attracts the <u>an</u>ions.

Sec. 19-2
page 482
Notice that the
symbol of emf in
the text is "ε"
rather than "E"
It is more
common to use
"E" for electric
field.

- **Electromotive force,** emf, E, of a source is the maximum energy per unit charge supplied by the source. If the circuit between cathode and anode is incomplete (open), the flow of current between them stops and the maximum potential difference, called the emf, develops across the cell. Therefore, V_{oc}, the potential difference or voltage across the cell when it is in an open circuit is the emf of the cell:

$$V_{oc} = E$$

The SI unit of emf is the volt, V.

Notice that "V" is used for the unit "vol" and the symbol representing the potential difference. This can be confusing.

The potential difference, V, across the source of emf in a closed circuit is less than the emf because of the internal resistance of the source:

$$E = V_{oc} > V$$

Sec. 19-2
page 483 of the
text gives a good
explanation of
internal
resistance and
terminal voltage.

Joule heating, Ir, is the heat produced by the collisions of the electrons with the nuclei and is equal to the product of the current, I, and the internal resistance, r. It occurs in all conductors, including the source of the current, the wires carrying the current and any devices in the circuit that use the current. The thermal energy increases the rate of vibration of the nuclei around their equilibrium positions in the solid.

Internal resistance, r, is the direct result of Joule heating. As the nuclear vibrations increase in frequency and length, the chances of colliding with the electrons increases. The more an electron collides with the nuclei, the slower its progress through the conductor. That is, the resistance to the flow of charge increases as the conductor heats up.

Terminal voltage: The irreversible or Joule heat produced by a current through a conductor produces the internal resistance. The potential difference across the electrodes or terminals of the cell is different from the emf of the cell:

$$V = E \pm Ir$$

The terminal voltage, V, equals the emf, E, plus or minus the Joule heat, Ir. The sign of Ir depends on the direction of current flow through the cell or battery.

1. If the cell is supplying the voltage, current flows through the circuit from its cathode to its anode. Then $V = E - Ir$, and the terminal voltage is less than the emf. This is generally the case for all cells and batteries.

2. If the cell is being "recharged," current is being forced to flow from the anode which is the reactant that gets reduced (gains electrons) to the cathode whose reactant is normally oxidized (loses electrons), then $V = E + Ir$, and the terminal voltage is greater than the emf.

3. If no current is flowing through the cell, no Joule heating can occur and $V = E$.

• A **battery** is a direct converter made up of two or more electrochemical cells connected to each other in series or in parallel or in a series-parallel combination. Batteries are therefore sources of direct current. The circuit diagram for a battery indicates the number of cells it contains. For example, the symbol for a three-cell battery is:

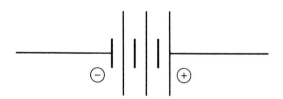

For a battery made of cells connected in series:

—The emf of the battery equals the sum of the emfs of the individual cells.
—The current in each cell of the battery and in the external circuit are equal in magnitude to the current in the battery as a whole.
—The internal resistance of the battery equals the sum of the internal resistances of the individual cells.

For a battery of identical cells connected in parallel:

—The emf of the battery and of each cell is constant and equal to the potential difference across the external circuit.

—The current in the circuit is equal to the current in the battery, which is the sum of the currents supplied by each cell.

—The reciprocal of the internal resistance of the battery is equal to the sum of the reciprocals of the internal resistance of each cell.

Sec. 18-3
page 463

D. Ohm's Law

Ohm's law of resistance states that in a closed circuit, the resistance, R, is inversely proportional to the current, I, and directly proportional to the voltage, V, of the circuit:

$$R = V/I$$

Therefore:

$$V = IR \text{ and } I = V/R$$

Ex. 18-2
page 464

Note that the voltage used in Ohm's law is the terminal voltage for the source in a closed circuit. This is less than the maximum emf which occurs in the open circuit.

The SI dimensions of the derived unit of resistance, the ohm, Ω are:

$$\Omega = V/A = (J/C)/(C/s) = (J\ s)/C^2 = (N\ m\ s)/C^2 = (kg\ m^2\ s)/(s^2\ C^2) = (kg\ m^2/C^2\ s)$$

1. For individual components of a circuit, R_i, V_i, and I_i are the variables.

2. Any set of components in a circuit can be replaced by a single equivalent component, and the symbols for the variables are R_{eq}, V_{eq}, and I_{eq}.

3. For the complete circuit, the total equivalent values are given as R_{tot}, V_{tot}, and I_{tot}.

E. Circuit Diagram

A circuit diagram shows the closed path available for the current to flow through. Electric current is a mechanism for transmitting energy. A closed conducting path or loop is required if the energy is to be used outside of the source. Devices or circuit components that can use the energy to perform work are called loads or resistances.

The current at any point in a circuit is measured with an ammeter; the ammeter is always connected in series with the device it is monitoring. The circuit diagram symbol is:

Ⓐ or Ⓘ

The voltage at any point in a circuit is measured with a voltmeter; the voltmeter is always connected in parallel with the device it is monitoring. The circuit diagram symbol is:

Ⓥ

• A **series circuit** has only one conducting path available to the current.

Sec. 19-1
page 480

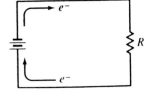

Figure 4.13-3

Typical series circuit.

1. Current: Since all current must go over the same path, its value is constant at all points in the circuit:

$$I_{tot} = I_i$$

The current flowing through any individual component, I_i, must be the same as the total current in the path.

2. Voltage: The total voltage in the circuit is equal to the applied emf (voltage source). This is equal to the algebraic sum of the individual voltage drops across each device in the circuit:

$$V_{tot} = emf = \Sigma\ V_i$$

3. Resistance: The total resistance in the circuit is the algebraic sum of the individual components:

$$R_{tot} = \Sigma\ R_i$$

Note: Devices in a series circuit should be rated to operate at the same current.

- A **parallel circuit** provides more than one conducting path. The circuit has two or more components connected across two common points in the circuit, and this provides the separate paths for the current.

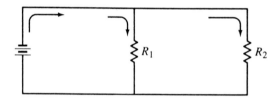

Figure 4.13-4

Typical parallel circuit.

1. Current: The total current is the sum of the currents in the individual branches and equal to the current supplied by the source:

$$I_{tot} = \Sigma\ I_i$$

2. Voltage: The potential difference is constant across all branches of the circuit:

$$V_{tot} = V_i$$

3. Resistance: The reciprocal of the total resistance is the algebraic sum of the reciprocals of the individual resistances:

$$1/R_{tot} = \Sigma\ 1/R_i$$

If only two resistances are connected in parallel, the equation can be rewritten as:

$$R_{tot} = (R_1R_2)/(R_1 + R_2)$$

The total resistance is the ratio of the product of the two resistances over their sum.

Note: Devices in a parallel circuit should be rated to operate at the same voltage.

- A **compound circuit** contains both series and parallel circuit sections. Each section can be resolved into its equivalent single R, V and I value.

Ex. 19-5 shows a
detailed analysis
of a compound
circuit.

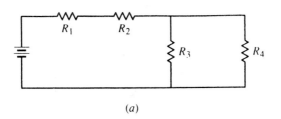

(a)

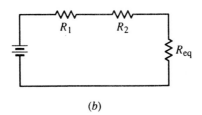

(b)

Figure 4.13-5
(a) Typical compound circuit. R_3 and R_4 are connected in parallel and can be replaced by the single equivalent resistance $R_{eq} = R_3 R_4/(R_3 + R_4)$. This resolves into circuit (b) where all resistances are connected in series so that $R_{tot} = R_1 + R_2 + R_{eq}$.

Sec. 19-3
page 485
"Junction rule"

- **Kirchhoff's laws:**

  1. Kirchhoff's first law or branch theorem states that the algebraic sum of the current entering any junction point in a circuit must be zero. It is based on the Law of Conservation of Charge.

  2. Kirchhoff's second law or loop theorem states that the algebraic sum of all changes in potential around a closed circuit is zero. It is based on the Law of Conservation of Energy.

F. Capacitor

Sec. 17-7
page 449

A **capacitor,** C, consists of two conductors (usually but not necessarily in the form of parallel plates) brought near to but not touching each other, separated by air, a vacuum, or an insulating material called a dielectric, and used to store electric current.

The space between the two plates means the circuit is open at that point. If the capacitor is connected to a potential difference, then opposite static charges will develop on the two plates. Since the charges are at rest, each plate is an equipotential surface and a uniform electric field exists in the space between them.

- **Capacitance** measures the ability of a capacitor to store electric charge. It is the ratio of the charge on either plate of a capacitor to the potential difference between them:

$$C = Q/V$$

The SI unit of capacitance is the farad, F, which equals one coulomb per volt:

$$1 \text{ F} = 1 \text{ C/V}$$

Capacitance is directly proportional to the area of each plate and inversely proportional to the distance separating them. For a parallel-plate capacitor, the

Ex. 17-6
page 450
Notice that the
text represents
(K/4πk) with
another constant
(ε₀) called the
"permittivity of
free space".

proportionality constant is $K/4\pi k$, where K is the dimensionless **dielectric constant** of the insulator and k is the constant in Coulomb's law:

$$C = \frac{K}{4\pi k}\left(\frac{A}{d}\right)$$

A large capacitance is produced by:

1. a large plate area.

2. an insulator with a high dielectric constant.

3. a small distance between the plates.

- A **dielectric** is the insulating material between the two plates of the conductor. The nature of the dielectric medium changes the capacitance.

The **dielectric constant, K,** is the ratio of the capacitance when the plates are separated by the insulator to the capacitance when they are separated by a vacuum:

$$K = C_{diel}/C_{vac}$$

K is a dimensionless number that ranges from 1 to 10 for most common dielectric materials.
The net effect of a dielectric between the plates is to lower the potential gradient of the electric field between the plates.

- Capacitors in circuits:

    —In series: The current is constant and the voltage drop across each capacitor is:

$$V_i = Q/C_i$$

Therefore,

$$V_{tot} = \Sigma\, V_i = Q\,\Sigma\, 1/C_i$$

And the reciprocal of the total capacitance is the sum of the reciprocals of the individual capacitances:

$$1/C_{tot} = \Sigma\, 1/C_i$$

    —In parallel: The voltage is constant and the individual charge on each capacitor is:

$$Q_i = C_i V$$

Therefore,

$$Q_{tot} = V\Sigma C_i$$

And the total capacitance is the sum of the individual capacitances:

$$C_{tot} = \Sigma C_i$$

Sec. 18-6
page 468

G. Electric Work, Power and Energy:

- **Work** is required to move a charge Q through a potential difference V. The amount of charge moved is given by the product of the current and the time:

$$W_{el} = E_{el} = QV = ItV$$

Joules = coulomb volts = ampere second volt

The work is equal to the amount of energy given up by the charge in moving through the circuit. From **Joule's law,** the heat produced in a conductor is directly proportional to the resistance, the square of the current and the time the current is maintained:

$$W_{el} = I^2 Rt$$

- **Power** is work per unit time:

$$P_{el} = W_{el}/t = IV$$

The equation gives the power dissipated or developed as a charge moves through a circuit.

Ex. 18-6 and 18-7 page 469 shows typical problems involving electrical power.

Using Ohm's Law, power can be expressed as:

$$P = IV = I^2 R = V^2/R$$

The unit of power is the Watt, W.

- **Electric energy** is the product of the power consumed by the device and the time during which the device operates:

$$E_{el} = Pt$$

The unit of electrical energy is usually given as kilowatt hours, kWh.

H. Alternating Current

In the United States, house current is 120 volts ac. The instantaneous or maximum voltage varies from +170 V to −170 V during each cycle.

The effective value equals the square root of the mean of the maximum value and is therefore called the root mean square or rms value. It is the maximum value divided by $2^{1/2}$.

For an ac source, the effective value is the number of amperes which for a given resistance produces heat at the same average rate as the same number of amperes of steady, direct current.

Ex. 18-9 page 472

- **Root mean square current,** I_{rms}, is the effective ac current:

$$I_{rms} = I_{max}/2^{1/2} = 0.202 \, I_{max}$$

- **Root mean square voltage,** V_{rms}, is the effective ac current:

$$V_{rms} = V_{max}/2^{1/2} = 0.202 \, V_{max}$$

Practice Problems

1. In a solid metal conductor, electric current is the movement of

    A. electrons only.
    B. protons only.
    C. nuclei.
    D. electrons and protons.

2. What is the current in the circuit shown?

    A. 0.5 A
    B. 2.0 A
    C. 6.0 A
    D. 12 A

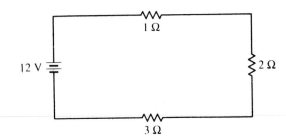

3. The ratio of the potential difference across a conductor and the current moving through the conductor is called the

    A. resistance.
    B. conductance.
    C. capacitance.
    D. electric potential.

4. For a circuit with constant resistance, which graph represents the relation between current and potential?

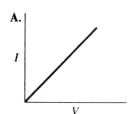 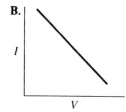

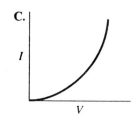 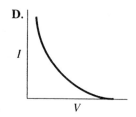

5. If the length of a conducting wire is doubled, the resistance of the wire will be

    A. quartered.
    B. halved.
    C. doubled.
    D. quadrupled.

Note: Problems 6 and 7 are based on the following circuit diagram:

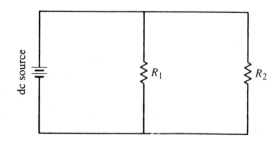

6. If $R_1 = 2.0\ \Omega$ and $R_2 = 6.0\ \Omega$, what is the total resistance of the circuit?

    A. 1.5 Ω
    B. 4.0 Ω
    C. 8.0 Ω
    D. 12 Ω

7. If the potential of a dc source is 5.0 volts, if $R_2 = 10\ \Omega$, what must be the potential difference across R_2?

    A. 0.50 V
    B. 2.0 V
    C. 5.0 V
    D. 50 V

8. Current in an ionized gas sample depends on

    A. cations only.
    B. anions only.
    C. free electrons only.
    D. cations, anions and free electrons.

9. The rate at which electrons pass a given point in a conductor is called the

    A. resistance.
    B. electric current.
    C. charge.
    D. potential difference.

10. If all the components of an electric circuit are connected in series, the physical quantity that is constant at all points in the circuit must be the

    A. voltage.
    B. current.
    C. resistance.
    D. power.

11. The current through a conductor is 3.0 A when it is attached across a potential of 6.0 V. How much power is used?

    A. 0.50 W
    B. 2.0 W
    C. 9.0 W
    D. 18 W

12. A 12 Ω load is connected across a 6.0 V dc source. How much energy does the load use in 1/2 an hour?

    A. 1.5×10^{-3} kWh
    B. 2.0×10^{-3} kWh
    C. 3.0×10^{-3} kWh
    D. 12×10^{-3} kWh

13. A current of 20.0 A flows through a battery with an emf of 6.20 V. If the internal resistance of the battery is 0.01 Ω, what is the terminal voltage?

    A. 1.24 V
    B. 6.00 V
    C. 6.40 V
    D. 31.0 V

14. Devices A and B are connected in parallel to a voltage source. If the resistance of A is four times as great as the resistance of B, then the current through A must be

    A. twice as great as that in B.
    B. half as great as that in B.
    C. four times as great as that in B.
    D. one fourth as great as that in B.

15. What happens to the resistance of a wire if its cross-sectional area is doubled?

    A. R is doubled.
    B. R is halved.
    C. R is quadrupled.
    D. R is quartered.

16. If the power produced by a circuit is tripled, the energy used by the circuit in 1 second will be

    A. multiplied by 3.
    B. divided by 3.
    C. multiplied by 9.
    D. divide by 9.

17. What must be the reading in the ammeter A for the circuit section shown?

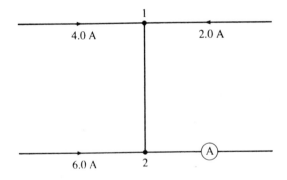

    A. 0 A
    B. 6.0 A
    C. 8.0 A
    D. 12 A

18. The two plates of a conductor are oppositely charged. What is the relationship among the intensities at the locations A, B and C?

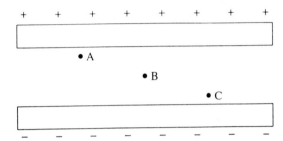

    A. Greater at A than at B
    B. Greater at B than at C
    C. Equal at A and C but less at B
    D. Equal at all three locations

19. What is the current in a wire if the flow of charge is 5.0 coulombs per 0.50 seconds?

    A. 1.0 A
    B. 2.5 A
    C. 5.0 A
    D. 10 A

20. For components connected in series, which quantity must be constant?

    A. Current
    B. Voltage
    C. Resistance
    D. Power

Answers and Explanations

1. **A** In a solid metal conductor, the positions of the atoms are fixed in the lattice and the electrons are free to move.

2. **B** The three resistors are connected in series so $R_{tot} = \Sigma R = 6\ \Omega$. From Ohm's Law, $V_{tot}/R_{tot} = I_{tot} = 12\ V/6\ \Omega = 2\ A$.

3. **A** Ohm's Law.

4. **A** Ohm's Law. V and I are directly proportional and linearly related. R is the slope of the curve.

5. **C** Resistance is directly proportional to length and inversely proportional to the cross-sectional area of the conductor. If the

area remains constant then doubling the length will double the resistance.

6. **A** The resistors are in parallel so:

$$R_{eq} = R_1R_2/(R_1 + R_2) = 12\ \Omega^2/8\ \Omega = 1.5\ \Omega$$

7. **C** For a parallel circuit, the potential difference is the same across all branches of the circuit.

8. **D** Conductivity in solids is due to free electrons; in electrolytic solutions it is due to ions; in ionized gases it is due to both ions and free electrons.

9. **B** Definition of current.

10. **B** A series circuit has only one path for current so it must be the same at all points in the circuit.

11. **D** Power = VI = 18 W

12. **A** The energy used by a load is the product of the power it uses per unit time and the length of time it is operated:

$$\begin{aligned} E &= W = Pt = V^2t/R \\ &= (6.0\text{ V})^2\ (0.50\text{ h})/12\ \Omega \\ &= 1.5 \times 10^{-3}\text{ kWh} \end{aligned}$$

13. **B** The Ir drop across the battery is the Joule heat produced:

$$Ir = (20\text{ A})(0.01\ \Omega) = 0.20\text{ V}$$

Since the battery is producing current and not being recharged, the terminal voltage will be less than the emf by the Joule heat produced. This eliminates choices C and D immediately:

$$V = E - Ir = 6.20\text{ V} - 0.20\text{ V} = 6.00\text{ V}$$

14. **D** Ohm's Law. For constant V, I and R are inversely proportional and linear. If R increases by four, then I must be divided by four.

15. **B** Resistance is inversely proportional to the area; double the area and you reduce the resistance by half.

16. **A** Power is the rate of using energy, $W = Pt$. Work and power are directly proportional and linear.

17. **D** Kirchhoff's law says the current entering a junction must equal the current leaving the junction. The current entering junction 1 is 6.0 A, so the current entering junction 2 is 12 A and the current leaving 2 and going through the ammeter is 12 A.

18. **D** The intensity of the electric field between the plates of a capacitor is uniform.

19. **D** Current is the rate of flow of charge past a given point in 1 second.

20. **A** Ohm's Law.

Key Words

ac root mean square current	dielectric constant	Joule's law
ac root mean square voltage	direct current	Kirchoff's laws
alternating current	drift velocity	laws of resistance
battery	electric energy	Ohm's law
capacitance	electric power	parallel circuit
capacitor	electric work	resistance
compound circuit	electrodes	resistivity
current	electromotive force	series circuit
dielectric	internal resistance	terminal voltage
		voltage

4.14 Atomic and Nuclear Structure

Sec. 30-1
page 800

A. Structure of the Nucleus

- The **atomic number,** Z, is the number of protons in the nucleus, which determines the identity and therefore the properties of the atom. All atoms of the same element have the same atomic number.

- The **mass number,** A, is the sum of the number of protons and neutrons in the nucleus of an atom. The neutrons affect the mass of the nucleus but not any nonmass-related properties.

Nucleons are the subatomic particles found in the nucleus: the protons and the neutrons. The number and ratio of the nucleons can only be changed by nuclear processes, not by general chemical reactions.

Note: Electrons are extranuclear particles because they are outside of the nucleus. The number of electrons associated with an atom can be changed readily by chemical processes.

- **Isotopes** are atoms of the same element that have different numbers of neutrons in the nucleus. Except for mass, the properties of isotopes of the same element are identical.

Nuclides are nuclear species. Isotopes are nuclides that have the same number of protons (atomic number) but different numbers of neutrons (and therefore, mass number).

The element symbol, $^A_Z X$, represents a given isotope of a given element, X. The superscript gives the mass number and the subscript the atomic number.

For example, the two isotopes of copper are $^{63}_{29}Cu$ and $^{65}_{29}Cu$. Each has 29 protons in the nucleus. ^{63}Cu also has 34 neutrons per nucleus giving it a total of $29 + 34 = 63$ nucleons. Isotope ^{65}Cu has two more neutrons than ^{63}Cu.

- The **atomic weight,** AW, of an element is the average atomic mass of the atoms of the element and reflects the relative abundance of the isotopes of the element.

Optional Aside: The nucleus contains almost all of the mass of the atom. The resting mass of an electron is $m_e = 9.11 \times 10^{-31}$ kg. It takes about 1876 electrons to equal the mass of one proton or neutron. Therefore, the contribution of electrons to the atomic weight is negligible. The heaviest known element is atomic number 109. Its 109 electrons are equivalent to less than 5 percent of the mass of a single nucleon.

also referred to as
"binding energy"
Ex. 30-1
page 803

Mass defect: The AW should equal the mass of all the protons and neutrons in the nucleus. The resting mass of a proton is $m_p = 1.6726 \times 10^{-27}$ kg and that of a neutron is $m_n = 1.6750 \times 10^{-27}$ kg.

$$AW = m_p(\text{number of protons}) + m_n(\text{number of neutrons})$$

The actual weight is generally less than this value. The difference is called the mass defect. The defect arises because energy must be expended by the nucleus to overcome the repulsion of the protons for each other. This energy is provided by converting some of the mass of the nucleus into energy. The transformation is given by Einstein's equation:

$$E = mc^2$$

Sec. 30-2
page 803
Intranuclear forces: Each nuclide is held together by three forces which act to overcome the repulsions due to the like nuclear charges.

1. The nuclear interaction is a very strong but short-ranged force that binds the nucleons together.

2. Electronic interaction is smaller in magnitude and increases in importance with increasing numbers of protons.

3. Weak interaction is the weakest of the three forces and is responsible for β decay nuclear processes.

Note: There is a fourth force, the gravitational interaction, which is the weakest force and not significant in nuclear physics.

- **Electron configuration:** In a neutral atom, the number of electrons circling the nucleus equals the number of protons in the nucleus.

Newtonian or classical physics states that a circulating charge continuously radiates energy. If this occurs for the electrons, they would eventually spiral into the nucleus as they lost energy and the ability to maintain their orbits.

However, quantum mechanics or modern physics states that an electron can move in an orbit at certain specific distances from the nucleus without losing energy. Since only certain distances are allowed, they are said to be quantized. If all values were allowed, the system would be continuous.

The radial distance of the orbital from the nucleus is given by the principle quantum number, n, which can have integer values from 1 to infinity with $n = 1$ representing the orbital closest to the nucleus.

For a one-electron system, each orbital distance has an energy associated with it, given by:

$$E_n = -Rhc/n^2$$

where R is a constant that depends on the identity of the nucleus; h is Planck's constant, 6.63×10^{-34} J s; and c is the constant speed of light in a vacuum, 3.00×10^8 m/s.

If all the electrons are in the lowest energy orbitals they can occupy, the atom is in its ground state, GS. Each element has one and only one ground state.

If the electrons are distributed in any other orbitals, the atom is in an excited state, ES. There are an infinite number of excited states possible for each element.

Sec. 27-9
page 733
Electronic transitions: For an electron to move from one allowed orbital to another, it must change its energy by an amount equal to the difference in energy between the two orbitals.

Absorption: For the electron to move from one orbital A to another orbital B that is higher in energy, the electron must absorb a quanta of energy in the form of a photon that is equal to the difference in energy between orbitals A and B:

$$\Delta E = E_{\text{final state}} - E_{\text{initial state}} > 0$$

Absorption is an excitation process. Excitation can occur from the GS to an ES, or from one ES to a higher ES. The energy absorbed is in the form of a photon:

$$E_{\text{photon}} = |\Delta E|$$

The pattern of photons absorbed is called the absorption spectrum of the atom.

Emission: The electron moves from an orbital that is higher in energy to one that is lower in energy. The electron must give up or emit a quanta of energy in the form of a photon that is equal to the difference in energy between the two orbitals. Emission is a relaxation process. The pattern of photons emitted is called the emission spectrum for the atom:

$$\Delta E = E_{\text{final state}} - E_{\text{initial state}} < 0$$

and the energy of the photon is again:

$$E_{\text{photon}} = |\Delta E|$$

An emission spectrum is not possible for an atom in its GS. The electron must first be excited into a higher energy orbital (absorbs energy).

Note: For transitions between the same two orbitals the quanta of energy and therefore the photon are identical for both absorption and emission. The energy of the photon is given by the absolute value for the energy change. The actual energy of a photon is always positive. The sign associated with ΔE simply tells you if the photon has been absorbed ($+$) or emitted ($-$).

Fluorescence and phosphorescence:

Sec. 28-10
page 766

Fluorescence is the emission of light from an atom as it relaxes from any ES. Fluorescence occurs as soon as excitation is over. Generally, the fluorescence radiation has a longer wavelength than the radiation originally used to excite the atom.

Phosphorescence is emission that occurs a measurable time interval after excitation has ended. This occurs because electrons in certain energy levels form metastable, long-lived states. On average, electrons stay in these energy levels for one second before relaxing.

Ionization potential is the amount of energy required to remove an electron completely from the atom. It is the difference between the energy of the orbital the electron occupies and the energy of the $n = \infty$ orbital. E_∞ always equals 0 J for all atoms. This is the orbital with the minimum distance required for the electron and nucleus to feel no coulombic attractive force.

For example: 13.60 eV of energy is enough to ionize a hydrogen atom. The ionization potential for removing the one electron is therefore 13.60 V.

B. Nuclear Properties

About 1500 nuclides are known. Of these, approximately 300 are stable. The remaining 1200 are unstable (all nuclei above $Z = 83$ are unstable). They are called radionuclides.

- Radionuclides are **radioactive.** They spontaneously undergo nuclear decay processes to form more stable nuclei.

   The decay processes can include emission or capture of particles or the emission of electromagnetic radiation.

Sec. 30-8
page 812

Half-life, $\tau_{1/2}$, is the time required for half of a sample of a given radionuclide to undergo radioactive decay.

Decay is a random process; therefore, it is not possible to predict if or when a given nuclide will decay. However, the half life gives the probability that any particle will decay. The original nuclide is called the parent, and the new nuclide formed after the decay process is called the daughter.

If T is the period of the half-life, then at some initial time $t = 0$ there will be N_0 nuclei. At time $t = T$ there will be only half as many of the original nuclei left, $N_0/2$. At t

= $2T$, there will be half of $N_0/2$ left, or $N_0/4$, etc. The change in the number of parent nuclei present, ΔN, in a given time interval Δt depends on the number of nuclei initially present, N_0, and the decay constant or Disintegration constant, λ:

$$\Delta N = -\lambda N_0 \Delta t$$

The negative sign appears because the number of nuclei is decreasing, so ΔN is always negative.

The equation can also be expressed as

$$\ln N - \ln N_0 = -\lambda t \quad \text{or} \quad N/N_0 = e^{-\lambda t}$$

where $e = 2.718$, the base of natural logarithms. The last equation has the advantage that it clearly shows radioactive decay as an exponential process.

The decay constant is related to the half-life by:

$$\lambda = \ln 2/\tau_{1/2} = 0.693/\tau_{1/2}$$

A short half-life means a large decay constant. λ has the units of reciprocal seconds, s^{-1}.

Ex. 30-4
page 813

The **activity** of a sample is the rate of decay, usually measured as disintegrations per second. It is the product of the decay constant and the number of nuclei present:

$$\text{activity} = \lambda N$$

The activity must decrease with time because N decreases with time.

- Conservation of energy and mass: The total number of particles (atomic number and mass number) must remain the same.

$$^{238}_{92}U \rightarrow {}^{234}_{90}Th + {}^{4}_{2}\alpha$$

The Uranium-238 isotope undergoes decay giving off the helium nucleus. The resulting element must contain the remaining protons and neutrons not emitted:

$$\text{Mass \#} = 238 - 4 = 234$$
$$\text{Atomic \#} = 92 - 2 = 90$$

Sec. 30-4
page 806

- α-Decay:

α-particles are bare helium nuclei, ${}^{4}_{2}He^{+2}$, or ${}^{4}_{2}\alpha$, with a net charge of positive 2.

In α-decay the parent nuclide reduces its nuclear size by two protons and two neutrons. The daughter is a new nuclide of a different element, having an atomic number two less than that of the parent. The element underwent a transmutation into a different element.

α-decay occurs in nuclides that are unstable because their nuclei are too large:

$$^{228}_{90}Th \rightarrow {}^{4}_{2}\alpha + {}^{224}_{88}Ra$$

Thorium is transmuted into radium. Conservation requires that the total number of protons and the total number of neutrons be identical on both sides.

Once formed, the α-particles readily react with any available electrons to produce neutral helium atoms.

α-particles have only short-range penetration and are easily stopped by a sheet of paper or by a few centimeters of air.

They are always emitted with the same kinetic energy.

Optional Aside: The strong interaction that holds nucleons together is short ranged, while the coulombic repulsion of protons toward each other has unlimited range (although it does drop off with the square of the distance between the interacting particles). As the size of the nucleus increases, the number of protons increases and the magnitude of the repulsive force becomes comparable and eventually exceeds the strong interaction. At this point, the nucleus is too large to be stable.

Sec. 30-5
page 808

- **β-Decay:**

β-particles are high-speed electrons, $_{-1}^{0}e-$ or $_{-1}^{0}\beta$.

They have greater penetration than α-particles and can penetrate thin metal sheets. They are stopped by a millimeter of lead or several centimeters of flesh.

β-particles are emitted with variable kinetic energies. In β-decay a neutron disintegrates into a proton that remains in the nucleus and an electron that is emitted as the β-particle. The resulting daughter has a Z that is increased by one and an A that remains the same (because a neutron is replaced by a proton so the total number of nucleons is unchanged).

$$_{6}^{14}C \rightarrow\ _{7}^{14}N +\ _{-1}^{0}\beta$$

β-decay occurs in nuclei that have too many neutrons relative to the number of protons present.

Positron emission:

This is an analog of β-decay that involves emission of the electron antiparticle, the positron $_{+1}^{0}e$.

Emission of a positron is accomplished by the conversion of a proton into a neutron. Z decreases by one and A remains unchanged.

$$_{7}^{13}N \rightarrow\ _{6}^{13}C +\ _{+1}^{0}e$$

This occurs in a nucleus that has too many protons relative to the number of neutrons present.

The neutrino, ν.

The kinetic energy of beta and positron particles occurs because both processes can also form a massless, uncharged particle called the neutrino. It functions to carry away energy without a change in mass.

Since the actual decay process can form two particles, the decay energy can be divided between the pair in any combination. Therefore, beta and positron particles have variable kinetic energies when emitted.

β-decay can occur for either a free neutron or for one in the nucleus:

$$_{0}^{1}n \rightarrow\ _{1}^{1}p +\ _{-1}^{0}e + \nu$$

Positron emission can occur only for protons in the nucleus but not for free protons:

$$_{1}^{1}p \rightarrow\ _{0}^{1}n +\ _{+1}^{0}e + \nu_{anti}$$

(The energy particle released is actually an antineutrino.)

Electron capture is the equivalent of positron emission. Z decreases by one and A remains the same. A proton captures an electron and forms a neutron and a neutrino:

$$_{1}^{1}p +\ _{-1}^{0}e \rightarrow\ _{0}^{1}n + \nu$$

Sec. 30-6
page 810

• **γ-decay:** γ-rays are high energy electromagnetic quanta or photons.

γ-rays are neutral and massless which make them similar to X-rays. However, their wavelengths are typically less than 1/100 those of X-rays.

The emission of γ-rays reduces the energy of the nucleus without changing the number of protons and neutrons. This is the only decay process discussed that does not result in a transmutation of the nuclide.

γ-rays are much higher in energy than either α- or β-particles and, therefore, have greater penetration.

• Fission and fusion reactions: The binding energy per nucleon that holds the nucleons together in the nucleus is greatest in intermediate-sized nuclei. Very light and very heavy nuclei have lower binding energies per nucleon ratios and are unstable.

In nuclear **fission**, a heavy nucleus splits into two lighter nuclei after absorbing a neutron. In the process several other neutrons are emitted so that a chain reaction is established:

$$^{235}_{92}U + ^{1}_{0}n \rightarrow ^{140}_{54}Xe + ^{94}_{38}Sr + 2\,^{1}_{0}n + \text{energy}$$

In nuclear **fusion,** two light nuclei fuse to form a single heavier nuclei. The binding energy per nucleon of the daughter is greater than that of the parents. The difference in binding energies is released in the reaction:

$$^{2}_{1}H + ^{3}_{1}H \rightarrow ^{4}_{2}He + ^{1}_{0}n + \text{energy}$$

Practice Problems

1. The main force responsible for holding the nucleons together in a nuclide is

   A. coulombic force.
   B. nuclear force.
   C. electronic force.
   D. gravitational force.

2. The energy of the $n = 2$ level of a particular element X is -3.6 eV and that of the $n = 4$ level is -0.9 eV. What is the energy change in going from the $n = 2$ to the $n = 4$ level?

   A. 0.25 eV
   B. 2.7 eV
   C. 4.0 eV
   D. -2.7 eV

Problems 3, 4, and 5 involve the conservation law discussed in Sec. 30-7 page 811.

3. In the nuclear equation below, X represents

   $$^{22}_{11}Na \rightarrow ^{22}_{10}Ne + ^{0}_{1}X$$

   A. an α-particle.
   B. a β-particle.
   C. a positron.
   D. a γ-photon.

4. What is the value of the mass number of the daughter nuclide in the equation below?

   $$^{34}_{15}P \rightarrow ^{A}_{Z}S + ^{0}_{-1}e$$

   A. 14
   B. 15
   C. 33
   D. 34

5. What is the atomic number of the daughter nuclide in the following reaction?

   $$^{30}_{15}P \rightarrow ^{A}_{Z}Si + ^{0}_{+1}e$$

   A. 14
   B. 16
   C. 30
   D. 31

6. If $^{13}_{7}N$ has a half-life of about 10.0 minutes, how long will it take for 20 grams of the isotope to decay to 2.5 grams?

   A. 5 min
   B. 10 min
   C. 20 min
   D. 30 min

7. A certain radionuclide decays by emitting an α-particle. What is the difference between the atomic numbers of the parent and the daughter nuclides?

A. 1
B. 2
C. 4
D. 6

8. What is the difference in mass number between the parent and daughter nuclides after a β-decay process?

A. -1
B. 0
C. 1
D. 2

9. Which species has no net charge?

A. α-particle
B. electron
C. proton
D. neutrino

10. A nitrogen atom has 7 protons and 6 neutrons. What is its atomic mass number?

A. 1
B. 6
C. 7
D. 13

11. Which is an isotope of $^{182}_{63}X$?

A. $^{182}_{62}X$

B. $^{182}_{64}X$

C. $^{180}_{63}X$

D. $^{180}_{62}X$

12. Which reaction is a fission reaction?

A. $^3_1H + ^1_1H \rightarrow ^4_2He + \text{energy}$

B. $^{35}_{17}Cl + ^1_1H \rightarrow ^{32}_{16}S + ^4_2He$

C. $^{235}_{92}U + ^1_0n \rightarrow ^{140}_{54}Xe + ^{94}_{38}Sr + 2\,^1_0n + \text{energy}$

D. $^9_4Be + ^4_2He \rightarrow ^1_0n + ^{12}_6C$

13. In the following equation, what is X?

$$^{14}_7N + ^4_2He \rightarrow ^{17}_8O + X$$

A. a proton
B. a positron
C. a β-particle
D. an α-particle

14. What is the number of neutrons in $^{140}_{54}Xe$?

A. 54
B. 86
C. 140
D. 194

15. Radon gas, Rn, has a half-life of 4 days. If a sample of Rn gas in a container is initially doubled, the half-life will be

A. halved.
B. doubled.
C. quartered.
D. unchanged.

16. A radionuclide decays completely. In the reaction flask only helium gas is found. The decay process was probably

A. β-decay.
B. α-decay.
C. γ-decay.
D. positron emission.

17. What is the half-life of a radionuclide if 1/16 of its initial mass is present after 2 hours?

A. 15 min
B. 30 min
C. 45 min
D. 60 min

18. The half-life of $^{22}_{11}Na$ is 2.6 years. If X grams of this sodium isotope are initially present, how much is left after 13 years?

A. $1/32X$
B. $1/13X$
C. $1/8X$
D. $1/5X$

Answers and Explanations

1. **B** Definition.

2. **B** $\Delta E = E_{final} - E_{initial} = E_4 - E_2 = (-0.9 \text{ eV})$

    $- (-3.6 \text{ eV}) = + 2.7 \text{ eV}$

3. **C** The superscript indicates $_1^0X$ is a massless particle. The subscript must give its charge. This fits the description of the positron which is the antiparticle of the electron.

4. **D** The mass number is A. Conservation of matter means:

$$34 = A + 0. \text{ Therefore, } A = 34$$

5. **A** The atomic number is Z. Conservation of charge means:

$$15 = Z + 1. \text{ Therefore, } Z = 14$$

6. **D** It will require several half-lives to get the amount involved. Choice A is eliminated because it is less than one half-life. Similarly, choice B is eliminated because it is exactly one half-life.

In every 10-minute period, half of the nuclei will decay and therefore, the mass will decrease to half its previous value. After the first 10 minutes only 10 grams will remain. After the second 10-minute interval, only 5 grams will remain. After the third, only 2.5 grams. It takes three 10-minute half-lives to reach 2.5 grams so the time required is 30 minutes.

7. **B** An α-particle is a $_2^4$He helium nucleus. In α-decay, two protons are effectively removed.

8. **B** β-decay emits a high energy electron, $_{-1}^0e$. In the process, a neutron decays into a proton plus the emitted electron. The number of nucleons remains unchanged with the proton replacing the neutron in the sum of nucleons.

9. **D** By definition.

10. **D** The mass number is the sum of neutrons and protons = 6 + 7 = 13 nucleons.

11. **C** Isotopes of an element have the same atomic number but different numbers of neutrons and thus, different mass numbers. The atomic number of X is 63.

12. **C** In a fission reaction a heavy nucleus absorbs a neutron and splits into two lighter weight nuclei. The process releases other neutrons and energy and can support a chain reaction.

13. **A** The conservation of matter means the mass number of X must be 1:

$$14 + 4 = 17 + Z. \text{ Therefore, } Z = 1$$

and the atomic number is 1:

$$7 + 2 = 8 + A. \text{ Therefore, } A = 1$$

This is the description of a proton, $_1^1p$.

14. **B** The number of neutrons is the mass number minus the atomic number: $\Sigma(n + p) - \Sigma p = 140 - 54 = 86$.

15. **D** The half-life is a constant that depends on the identity of the nuclide, not on the amount of nuclide present.

16. **B** The α-particle is the helium nucleus. Each α-particle then acquires two electrons to form a neutral helium atom.

17. **B** In every half-life, the mass decreases to half its previous value:

$$1/16 = 1/2^4$$

It takes four half-lives to decay down to 1/16 the original mass. Each must be 30 minutes long since the entire process takes two hours.

18. **A** In 13 years, there will be 5 half-lives of 2.6 years each ($5 \times 2.6 = 13$). The mass decreases to $1/2^5 = 1/32$ its original value.

Key Words

activity	fission	mass defect
α decay	fluorescence	mass number
atomic number	fusion	neutrino
atomic weight	γ decay	nucleons
β decay	half-life	nuclides
electron capture	intranuclear forces	phosphorescence
electron configuration	ionization potential	positron emission
electronic transitions	isotopes	radioactive

Chapter 5

Physical Sciences Practice Exam

Time: 100 minutes
77 Questions

DIRECTIONS: This test includes 10 sets of questions related to specific passages (62 questions) and 15 shorter, independent questions. The passages represent all four format styles that you may see on the MCAT (Information Presentation, Problem Solving, Research Study and Persuasive Argument). Explanatory answers immediately follow the test.

Passage I (Questions 1–5)

Theory 1. Early in the twentieth century, many chemists believed that the stability of the methane molecule, CH_4, could be explained by the "octet" rule, which states that stability occurs when the central atom, in this case carbon, is surrounded by 8 "valence," or outer, electrons. Four of these originally came from the outer electrons of the carbon atom itself, and four came from the four surrounding hydrogen atoms (hydrogen was considered to be an exception to the rule, since it was known to favor a closed shell of two electrons, as helium has).

According to the octet rule, neither CH_3 nor CH_5 should exist as stable compounds, and this prediction has been born out by experiment.

Theory 2. While the octet rule predicted many compounds accurately, it also had shortcomings. The compound PCl_5, for example, is surrounded by 10 electrons. The greatest shock to the octet rule concerned the "noble gases" such as krypton and xenon, which have eight electrons in their atomic state and therefore should not form compounds, since no more electrons are needed to make an octet. The discovery in 1960 that xenon could form compounds such as XeF_4 forced consideration of a new theory, which held that (a) compounds can form when electrons are completely paired, either in bonds or in nonbonded pairs; (b) the total number of shared electrons around a central atom can vary and can be as high as 12; and (c) the

153

shapes of compounds are such as to keep pairs of electrons as far from each other as possible.

For example, since sulfur is surrounded by 6 electrons in the atomic state, in the compound SF_6 it acquires 6 additional shared electrons from the surrounding fluorines for a total of 12 electrons. The shape of the compound is "octahedral," as shown below, since this configuration minimizes the overlap of bonding pairs of electrons.

F
F | F
  \ | /
    S
  / | \
F | F
    F

1. According to Theory 1, the compound CH_2Cl_2:

A. should have 8 electrons surrounding the carbon atom.

B. cannot exist since the original carbon atom does not have 8 electrons.

C. should have 8 electrons surrounding each hydrogen atom.

D. requires more electrons for stability.

2. According to Theory 1, the compound XeF_4:

A. exists with an octet structure around the xenon.

B. should not exist since the xenon is surrounded by more than eight electrons.

C. will have similar chemical properties to CH_4.

D. exists with the xenon surrounded by 12 electrons.

3. The atom boron has three outer electrons. In bonding to boron a fluorine atom donates one electron. The BF_3 molecule is known to exist. Which of the following is true?

A. BF_3 is consistent with Theory 1.

B. The existence of BF_3 contradicts Theory 2.

C. According to Theory 2, the structure of BF_3 is a pyramid:

D. According to Theory 2, the structure of BF_3 is triangular and planar:

F
|
B
/ \
F   F

4. A scientist seeking to explain why Theory 2 has more predictive power than Theory 1 might argue that:

A. 8 electrons represent a "closed shell."

B. while 8 electrons represent a "closed shell" for some atoms, for others the closed shell may be 6, 10, or 12.

C. it is incorrect to assume that a given atom always has the same number of electrons around it.

D. CH_4 is not as important a compound as XeF_4

5. Theory 2 could be undermined by evidence:

A. of the existence of SF_4.

B. of the existence of XeF_5.

C. showing that CH_4 was more stable that XeF_4.

D. of molecules with stable octets.

Passage II (Questions 6–13)

Among the simplest molecules are those formed from two identical atoms in the first rows of the periodic table. Bond strengths and bond lengths in these molecules vary considerably but can be predicted from molecular orbital theory.

Table 1 gives properties of several homonuclear diatomic molecules and molecular ions. A dash (—) indicates that a species does not exist as a stable entity.

Table 1

Molecule	Bond length (Å)	Bond energy (kJ/mole)
H_2^+	1.06	256
H_2	0.741	432
He_2^+	1.08	322
He_2	—	—
Li_2	2.67	110
B_2	1.59	274
C_2	1.24	603
N_2	1.10	942
O_2	1.21	494
Cl_2	1.99	239
Cl_2^+	1.89	415
Ne_2	—	—

Figure A shows the energy levels in these molecules.

Figure A: Molecular Orbital Diagram

$\overline{\sigma^*}(2p)$		$\overline{\sigma^*}(2p)$
$\overline{\pi_x^*}\ \overline{\pi_y^*}$		$\overline{\pi_x^*}\ \overline{\pi_y^*}$
$\overline{\sigma}(2p)$		$\overline{\pi_x}\ \overline{\pi_y}$
$\overline{\pi_x}\ \overline{\pi_y}$		$\overline{\sigma}(2p)$
$\overline{\sigma^*}(2s)$		$\overline{\sigma^*}(2s)$
$\overline{\sigma}(2s)$		$\overline{\sigma}(2s)$
(a) Li_2 through N_2		(b) O_2 and F_2

6. According to Table 1, for how many stable, neutral molecules is the bond length under 1.50Å and the bond energy greater than 450 kJ/mol?

   A. 0
   B. 1
   C. 2
   D. 3

7. Which of the following trends does Table 1 illustrate?

   A. Bond length increases with molecular weight.

   B. Within a period, bond energy increases along the period.
   C. Within a period, bond length varies inversely with bond energy.
   D. Atoms that fail to form diatomic molecules have an odd number of valence electrons.

8. Consider the neutral molecule/positive ion pairs shown in Table 1, which statement is generally true?

   A. removing an electron from a molecule X_2 to form X_2^+ lowers both the bond energy and the bond length.
   B. removing an electron from a molecule X_2 to form X_2^+ lowers the bond energy and increases the bond length.
   C. removing an electron from a molecule X_2 to form X_2^+ raises the bond energy and decreases the bond length.
   D. None of the above.

9. Suppose that the bonding in N_2 is represented by the Lewis structure

   :N:::N:

   and in O_2 by the structure

   :Ö::Ö:

   Suppose also that triple bonds as shown in Lewis structures can be taken to be shorter than double bonds. Then the Lewis structures are:

   A. consistent with the data shown in the table.
   B. consistent with the bond energy data, but inconsistent with the bond length data.
   C. consistent with the bond length data, but inconsistent with the bond energy data.
   D. inconsistent with both the bond length data and the bond energy data.

10. According to Figure A, the molecular orbital diagram, what is the bond order of Li_2?

    A. 0
    B. 1/2
    C. 1
    D. 2

11. According to Figure A, why does He_2 not exist as a stable molecule?

    A. Its bond order is only 1/2.
    B. It would have a bond order of 0.
    C. It has no bonding electrons in orbitals deriving from atomic $2s$ orbitals.
    D. A stable molecule may not use σ^* orbitals for bonding.

12. Table 1 showed instances where the addition of electrons to a molecule resulted in decreased stability. Figure A shows that for the example of N_2 to O_2, the effect is explained as follows:

    A. The extra electrons are placed in bonding orbitals.
    B. The extra electrons are placed in antibonding orbitals.
    C. The extra electrons go into a combination of bonding and antibonding orbitals.
    D. Since electrons repel one another, adding them to a molecule invariably is destabilizing.

13. The trend by which bond length appears to correlate inversely with bond energy might be explained by noting which of the following?

    A. Higher-energy electrons prefer to be close together.
    B. Higher bond energy corresponds to more electrons, and each electron requires a given amount of space.
    C. Since higher bond energy corresponds to a greater number of electrons in the space between the nuclei, these electrons will tend to pull the nuclei toward each other.
    D. Since higher bond energy corresponds to a greater number of electrons in the space between the nuclei, these electrons will tend to pull the nuclei away from each other.

Passage III (Questions 14–19)

By formally defining "half reactions" we can predict the spontaneity of actual reactions which can be thought of as proceeding through a pair of half reactions. The table below gives the standard reduction potential of selected half reactions. Half reactions are arranged with the most spontaneous first.

Reaction	$E°$ (volts)
$F_2 + 2e^- \rightarrow 2F^-$	2.87
$Cr_2O_7^{2-} + 14H^+ + 6e^- \rightarrow 2Cr^{3+} + 7H_2O$	1.33
$Br_2 + 2e^- \rightarrow 2Br^-$	1.09
$Ag^+ + e^- \rightarrow Ag$	0.80
$Fe^{3+} + e^- \rightarrow Fe^{2+}$	0.77
$Cu^{2+} + 2e^- \rightarrow Cu$	0.34
$Cu^{2+} + e^- \rightarrow Cu^+$	0.15
$H^+ + e^- \rightarrow 1/2\ H_2$	0.00
$Pb^{2+} + 2e^- \rightarrow Pb$	−0.13
$Ni^{2+} + 2e^- \rightarrow Ni$	−0.25
$Zn^{2+} + 2e^- \rightarrow Zn$	−0.76

14. For how many of the reactions above does the absolute value of $E°$ decrease by more than 50% as one goes to the next reaction down on the list?

    A. 0
    B. 1
    C. 2
    D. 3

15. How many cells with a measured voltage greater than 1.5 v can be made using standard half cells with the half reactions listed?

    A. 6
    B. 13
    C. 15
    D. 17

16. Suppose that under a new system of reporting half-potentials, the value of 0.00 is to be assigned to the reduction of cupric ion to copper metal. Under this system, $E°$ for the reduction of F_2 would have what value?

    A. 2.53 v
    B. 2.72 v
    C. 3.02 v
    D. 3.21 v

17. Using values in the table, calculate $E°$ for the reaction $Pb^{2+} + Cu \rightarrow Pb + Cu^{2+}$.

   A. 0.21 v
   B. −0.21 v
   C. 0.47 v
   D. −0.47 v

18. A student finds an old chemistry textbook that expresses half-potentials as *oxidation* potentials; e.g., potentials for a reaction such as $Fe \rightarrow Fe^{2+} + 2e^-$. If the reduction potential table above were converted to an oxidation potential table, with the highest potentials on top, which of the following would be true?

   A. The F_2/F^- reaction would be listed first.
   B. The Cu^{2+}/Cu reaction would be above the Cu^{2+}/Cu^+ reaction.
   C. The Zn^{2+}/Zn reaction would be above the H^+/H_2 reaction.
   D. The Ni^{2+}/Ni reaction would be below the H^+/H_2 reaction.

19. Based on the information in the reduction potential table, a spontaneous reaction will occur when:

   I. copper metal is dropped into HNO_3.
   II. metallic lead is dropped into HNO_3.
   III. bromine gas is bubbled through a solution of $CuNO_3$.

   A. I only
   B. II only
   C. III only
   D. I and II only

Passage IV (Questions 20–25)

   While researching the solubility of various lead salts, an investigator decides to try a new way to graph the results. Although a chart of anion concentration against cation concentration usually shows a hyperbola or a similar curve, her choice of axes produces the straight lines shown below:

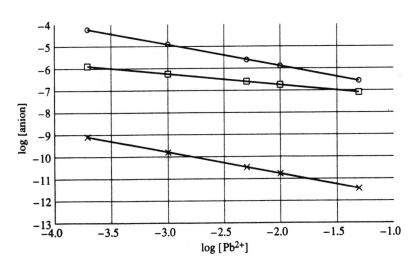

Key to anions:
 $-\!\!\circ\!\!-[SO_4^{2-}]$ $-\!\!\boxminus\!\!-[OH^-]$ $\!\!\times\!\!-[CO_3^{2-}]$

Solubilities of lead salts.

20. For a given concentration of lead ion, which anion has the lowest concentration?

    A. SO_4^{2-}
    B. OH^-
    C. CO_3^{2-}
    D. Cannot be determined

21. The absolute value of $\Delta\log$ (anion)/$\Delta\log$ [Pb^{2+}] is:

    I.   the same for SO_4^{2-} and OH^-.
    II. about twice as great for OH^- as for SO_4^{2-}
    III. about the same for SO_4^{2-} and CO_3^{2-}

    A. I only
    B. II only
    C. III only
    D. I and II only

22. Since the lowest curve is a straight line, we may say that:

    A. $[CO_3^{2-}]$ decreases linearly as [Pb^{2+}] decreases.
    B. $[CO_3^{2-}]$ increases linearly as [Pb^{2+}] increases.
    C. $\log [CO_3^{2-}]$ decreases linearly as $\log$ [Pb^{2+}] decreases.
    D. $\log [CO_3^{2-}]$ increases linearly as $\log$ [Pb^{2+}] decreases.

23. The slope of the lowest curve is closest to which of the following?

    A. $-1/2$
    B. $1/2$
    C. -1
    D. 1

24. Which of the following might explain the difference in the slopes of the top two curves?

    A. K_{sp} is higher for $PbSO_4$ than for $Pb(OH)_2$.
    B. The fact that there are 2 OH^-'s to every Pb^{2+} leads to a different graph for $Pb(OH)_2$ than for $PbSO_4$.
    C. $PbSO_4$ is more soluble than $Pb(OH)_2$.
    D. None of the above.

25. Since for $Pb(OH)_2$, $K_{sp} = [Pb^{2+}][OH^-]^2$,

    A. $\log K_{sp} = (\log [Pb^{2+}])(\log [OH^-])^2$
    B. $\log K_{sp} = (\log [Pb^{2+}])/(\log [OH^-])^2$
    C. $\log K_{sp} = \log [Pb^{2+}] + 2\log [OH^-]$
    D. $\log K_{sp} = (\log [Pb^{2+}])^3$

Passage V (Questions 26–32)

The study of equilibrium allows us to determine whether a reaction will proceed. The calculation uses the equilibrium constant, which can be derived from the standard free energy change. The study of reaction rates is essential if we are to determine the speed with which a reaction proceeds.

A chemist *predicts* that a reaction in which reactant A is converted to products B and C will proceed according to Figure 1.

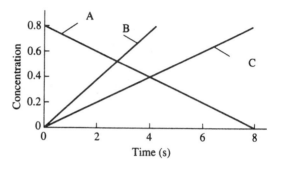

Figure 1

The chemist then *measures* the concentrations of the three species against time, and obtains the results shown in Figure 2.

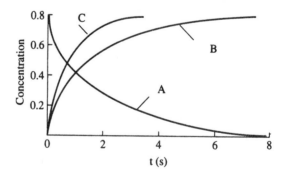

Figure 2

In order to determine the rate law for the reaction, the chemist replots the data as shown in Figure 3.

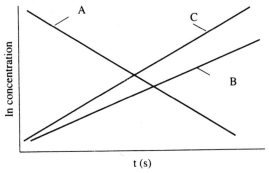

Figure 3

26. Using Figure 1, when $P_A = P_B$, what does P_C approximately equal?

A. $.5P_A$
B. P_A
C. $P_A - P_B$
D. $P_A + P_B$

27. As the reaction in Figure 1 proceeds, the total pressure of all species:

A. decreases.
B. stays the same.
C. increases.
D. increases, then decreases.

28. Which of the following summarizes the reaction graphed in Figure 1?

A. $A + B \rightarrow C$
B. $A \rightarrow B + C$
C. $A \rightarrow 2B + C$
D. $A \rightarrow B + 2C$

29. Which of the following expresses the rate of the reaction graphed in Figure 1?

A. $-dP_A/dt = k$
B. $-dP_A/dt = k[A]$
C. $-dP_A/dt = k[B]$
D. $-dP_A/dt = k[B][C]$

30. Compared to Figure 1, which predicted the concentrations over time, the initial rate of appearance of B in Figure 2 is:

A. lower.
B. the same.
C. higher.
D. sometimes lower, sometimes higher.

31. Based on Figure 2, the total pressure at 1 s is closest to:

A. 0.3
B. 0.5
C. 1.3
D. 1.8

32. Which of the following best describes the fit of the data in Figure 3 to the chemist's initial prediction?

A. the predicted kinetics were 0th-order; the experimental kinetics were in agreement.
B. the predicted kinetics were first-order; the experimental kinetics were in agreement.
C. the predicted kinetics were 0th-order; the experimental kinetics were first-order.
D. the predicted kinetics were first-order; the experimental kinetics were second-order.

Passage VI (Questions 33–39)

Reaction 1

An organic chemist obtains the kinetic data shown for the following reaction, where S is a "substrate," N a nucleophile, and X a "leaving group."

$$SX + N \rightarrow SN + X$$

[SX]	[N]	relative rate
.05	.05	1.20
.05	.10	2.40
.10	.05	2.40
.10	.10	4.80
.15	.15	10.80

Reaction 2

Next, the chemist obtains data for a second reaction:

$$S'X' + N' \rightarrow S'N' + X'$$

[S'X']	[N']	rel. rate
.005	.10	16
.005	.15	16
.0075	.5	24
.015	.30	48

Finally, the chemist investigates the effect of the nucleophile for each of the two reactions. Her results follow:

nucleophile	rate of Reaction 1	rate of Reaction 2
CH_3OH	very slow	Changing nucleo-
H_2O		philes had no
Cl^-	↓	effect on rate
NH_3	moderate	
Br^-		
OH^-	↓	
I^-		
SH^-	very fast	

33. In Reaction 1, the relative rate varies:

    I. linearly with the concentration of SX.

    II. linearly with the concentration of the nucleophile.

    III. inversely with SX.

  A. I only
  B. II only
  C. III only
  D. I and II only

34. In Reaction 2, increasing the concentration of N' while holding S' X' constant:

  A. can either increase or decrease the relative rate
  B. increases the relative rate
  C. decreases the relative rate
  D. has no effect on the relative rate

35. The rate of Reaction 1 is best expressed by which of the following?

  A. rel. rate = .05[SX] + .05[N]
  B. rel. rate = .05[SX]
  C. rel. rate = 480[SX]
  D. rel. rate = 480[N][SX]

36. The rate of reaction 2 is best expressed by which of the following?

  A. rel. rate = .005[S'X'] + .10[N']
  B. rel. rate = .005[S'X']
  C. rel. rate = 1600[S'X']
  D. rel. rate = 3200[S'X']

37. The researcher finds that for Reaction 3:

$$CH_3Cl + I^- \rightarrow CH_3I + Cl^-$$

the kinetics data resemble those for Reaction 1, whereas for Reaction 4:

$$(CH_3)_3\!-\!C\!-\!Cl + I^- \rightarrow (CH_3)_3\!-\!C\!-\!I + Cl^-$$

the kinetics data resemble those for Reaction 2. This result is best explained by which of the following?

  A. Reaction 2 is favored by a stable carbocation, (as in Reaction 4) while Reaction 1 is favored by attack of the nucleophile on the substrate (as in Reaction 3).
  B. Reaction 1 is favored by a stable carbocation (as in Reaction 3), while Reaction 2 is favored by accessible attack of the nucleophile on the substrate (as in Reaction 4).
  C. Reaction 2, a particular example of Reaction 3, is favored by the fact that CH_3 is a better leaving group than $C(CH_3)$.
  D. Stereo inversion is more likely in Reaction 4.

38. The chemist decides that "better" nucleophiles (i.e., those which enhance the rate):

    I. tend to be found in the conjugate acid form of a species rather than the conjugate base.

II. have a central atom which is located toward the lower end of a group (c) in the Periodic Table.

III. have a central atom located to the right side of the Periodic Table.

A. I only
B. II only
C. III only
D. II and III only

39. What do the results for Reaction 2 suggest?

A. There is no nucleophilic attack in Reaction 2.
B. The leaving group effects predominate over the nucleophilic effects.
C. The nucleophile plays no part in the rate-determining step.
D. Reaction 2 proceeds in a rapid, one-step manner.

Passage VII (Questions 40–43)

A student draws the following graphs based on different sets of data. The graphs drawn are:

ec. 2-11
page 34

40. The graph that shows a body in equilibrium, with the ordinate representing velocity and the abscissa representing time, is:

A. Graph A.
B. Graph B.
C. Graph C.
D. Graph D.

41. The graph that has axes representing velocity and time and that represents the motion with the largest acceleration is:

A. Graph A.
B. Graph B.
C. Graph C.
D. Graph D.

42. Which graph best represents a motionless body?

A. Graph A with axes of distance vs time
B. Graph A with axes of velocity vs time
C. Graph B with axes of acceleration vs time
D. Graph B with axes of velocity vs time

43. If all the graphs have axes of distance vs time, which one represents motion with the smallest constant nonzero velocity?

A. Graph A
B. Graph B
C. Graph C
D. Graph D

Passage VIII (Questions 44–47)

A 700 N person pushes a 900 N box 5.00 meters up a 15.0-meter-long ramp in 3.00 seconds. At that point on the ramp the box is 3.00 meters above the ground. The frictional force for the box on the ramp is 100 N.

Ch. 6
page 124

44. How much power was required to move the box?

A. 1167 W
B. 1500 W
C. 500 W
D. 584 W

Sec. 6-10
page 141

45. What is the potential energy gained by the box?

A. 2.70×10^3 J
B. 1.50×10^4 J
C. 2.65×10^4 J
D. 3.54×10^4 J

Sec. 6-4
page 131

46. What is the minimum force the person must apply to the box to get it moving?

figure 4-30
page 88

  A. 100 N
  B. 900 N
  C. 540 N
  D. 640 N

47. What is the minimum force required to keep the box where it is without sliding down the ramp?

  A. 100 N
  B. 640 N
  C. 540 N
  D. 440 N

Passage IX (Questions 48–51)

A student wants to make a musical instrument using a hollow tube and a glass of water. She tests her idea on a single unit. The tube is suspended in the water in the glass. A portion of the tube, length X, is above the water line. The remaining portion, length Y, is below the water line.

figure 12-12
page 318

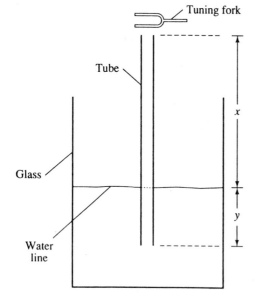

She strikes a tuning fork, holds it over the mouth of the tube and adjusts the tube's height until the sound is loudest.

48. What is the wavelength of the fundamental frequency of the sound produced by the tube?

  A. $2X$
  B. $4X$
  C. $2Y$
  D. $4Y$

49. When the tube is adjusted to make the sound louder, the wave property illustrated is:

  A. polarization.
  B. beats.
  C. resonance.
  D. refraction.

50. The wavelength of the sound that is refracted into the water at the air-water boundary is:

  A. less than $2Y$.
  B. greater than $4X$.
  C. between $2X$ and $4X$.
  D. exactly $4Y$.

51. The sound wave created by the tuning fork has a wavelength equal to which of the following?

  A. one complete period of vibration for the tuning fork
  B. one half the period of vibration for the tuning fork
  C. one quarter the period of vibration for the tuning fork.
  D. twice the period of vibration for the tuning fork.

Questions 52–77 are independent of any passage and independent of each other.

52. A student draws four small-diameter tubes labeled A through D placed in a reservoir of water.

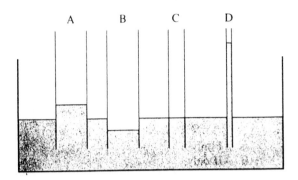

figure 10-33
page 264

Which two tubes in the drawing correctly illustrate the expected behavior of water?

A. A and D
B. B and C
C. A and C
D. C and D

53. A musician produces a certain note on a flute. If the musician blows harder, what happens to the sound wave associated with the note?

A. Its speed increases.
B. Its frequency increases.
C. Its frequency increases.
D. Its amplitude increases.

54. How long will it take sound to reach an observer who is 2200 feet away from the source if the speed of sound in air is approximately 1100 feet per second?

A. 1/2 second
B. 1 second
C. 2 seconds
D. 4 seconds

55. A ball is thrown vertically upward with a velocity of 40.0 m/s. What will be the height of the ball after 5.00 seconds?

A. 8.00 m
B. 77.5 m
C. 200 m
D. 322 m

56. Three children weighing 40 lb, 60 lb, and 80 lb respectively want to balance a 14-foot-long see-saw, pivoted at its center. The 80-lb child sits 2 ft from the left end of the see-saw. The 60-lb child sits 2 ft from the right end of the see-saw. Where must the 40-lb child sit?

A. 1.0 ft to left of pivot
B. 2.5 ft to right of pivot
C. 2.5 ft to left of pivot
D. 5.0 ft to right of pivot

Ex. 9-4
page 209

57. A 50 N box is set in motion. The coefficient of static friction is 0.50 and the coefficient of kinetic friction is 0.20. What is the difference between the force needed to initiate motion and the force needed to maintain motion?

A. 10 N
B. 15 N
C. 25 N
D. 125 N

Ex. 4-12
page 85

58. If air pressure is 14.7 lb/in^2, what is the total force exerted on one side of an 8 1/2 × 11 inch sheet of paper?

A. 63.6 lb
B. 137 lb
C. 6.36 × 10^3 lb
D. 1.37 × 10^3 lb

59. What is the electric current produced when a charge of 120 coulombs flows through a wire with a diameter of 2 cm in one minute?

A. 1 ampere
B. 2 amperes
C. 3 amperes
D. 4 amperes

Recall the definition of electric current.

60. The iodine isotope $^{131}_{53}$I decays to form the Xenon isotope $^{131}_{54}$Xe. What else results from this reaction?

A. An α-particle is emitted.
B. A positron is absorbed.
C. A positron is emitted.
D. A β-particle is emitted.

Consider the conservation of charge.

61. A pendulum of length L has a period of 0.5 second. To increase the period to 1.0 second, the length of the pendulum must be:

Eq. 11-8a
page 283

   A. doubled.
   B. quadrupled.
   C. halved.
   D. quartered.

62. The effect of β-particle emission is to produce a daughter nuclide that has:

Consider the
conservation of
charge.

   A. increased its atomic number by one.
   B. decreased its atomic number by one.
   C. increased its mass number by one.
   D. decreased its mass number by one.

63. Which of the following is not a correct empirical formula?

   A. $Ca_3(SO_4)_2$
   B. C_6H_{12}
   C. Na_2S
   D. N_2O_5

64. For those atoms in the first column of the periodic table which form stable positive ions, the ion usually has a considerably smaller radius than the neutral atom. What explains this effect?

   A. The outer electron(s) effectively screen the nuclear charge.
   B. The ion has a smaller nuclear charge.
   C. The neutral atom has a greater nuclear charge.
   D. The neutral atom has an outer electron in an orbital of higher "n" than the ion.

65. A gaseous air pollutant is found to have a mole fraction of 3×10^{-6} in the atmosphere. What is its partial pressure in atm if the barometric pressure is 1.00 atm?

   A. 10^{-6}
   B. 3×10^{-6}
   C. 6×10^{-6}
   D. $760 \times (3 \times 10^{-6})$

66. Which of the following results in a spontaneous reaction in water?

   I. addition of a weak base to a strong base
   II. addition of a weak acid to a strong base
   III. addition of a weak base to a strong acid

   A. I only
   B. II only
   C. III only
   D. II and III only

67. A reaction is found to have an equilibrium constant of 0.06 at a temperature of 298K and a pressure of 1.0 atm. It can be concluded that at these conditions,

   I. $\Delta H° > 0$
   II. $\Delta S° > 0$
   III. $\Delta G° > 0$

   A. I only
   B. II only
   C. III only
   D. I and II only

68. In the following reaction, which is (are) true?

$$H_2S + 2NO_3^- + 2H^+ \rightarrow S + 2NO_2 + 2H_2O$$

   I. H_2S is oxidized.
   II. H_2S is the oxidizing agent.
   III. NO_3^- is reduced.

   A. I only
   B. II only
   C. III only
   D. I and III only

69. A compound contains 5.9% hydrogen and 94.1% oxygen. What is its empirical formula?

   A. HO
   B. H_2O
   C. H_2O_2
   D. H_6O_{94}

70. The reaction

$$Ag^+ + Fe^{2+} \rightarrow Ag + Fe^{3+}$$

proceeds spontaneously. If a galvanic cell is constructed with Ag^+/Ag as one half cell and Fe^{3+}/Fe^{2+} as the other, the silver electrode will be:

A. the cathode, positive.
B. the cathode, negative.
C. the anode, positive.
D. the anode, negative.

71. The nuclear reaction:

$$^2_1H + ^3_1H \rightarrow ^4_2He + ^1_0n + \text{energy}$$

is an example of what type of reaction?

A. α-decay
B. fission
C. fusion
D. β-decay

Sec. 24-3
page 624

72. Monochromatic light passes through two parallel slits in a screen and falls on a piece of film. The pattern produced is an example of:

A. interference and reflection.
B. interference and diffraction.
C. refraction and diffraction.
D. diffraction and polarization.

73. Which of the following distances is shortest?

A. 0.456×10^4 cm
B. 4.56×10^4 m
C. 0.456 km
D. 4.56×10^6 mm

Sec. 6-4
page 131

74. A 70 kg man exerts a force of 20 N parallel to the surface of an incline while pushing a 50 N crate 10 meters up the incline. The incline is not frictionless. Which statement best describes the change in potential energy of the crate?

A. The E_p of the crate increases by less than 200 J.
B. The E_p of the crate increases by more than 200 J.
C. The E_p of the crate increases by less than 500 J.
D. The E_p of the crate decreases by less than 500 J.

75. Which of the following statements is most accurate for the system of object and concave mirror illustrated below?

figure 23-14
page 599

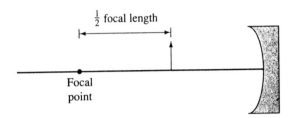

$\frac{1}{2}$ focal length

Focal
point

A. The image is larger than the object, inverted and real.
B. The image is larger than the object, upright and virtual.
C. The image is smaller than the object, upright and real.
D. The image is smaller than the object, inverted and virtual.

76. A 50 kg probe lands on a planet that has a mass that is twice as great as that of Earth and a radius that is twice as large as that of Earth. What is the gravitational force experienced by the probe?

A. The same as on the earth.
B. Half that on earth.
C. Twice that on earth.
D. Four times that on earth.

77. What happens to the kinetic energy of a body if its speed is tripled?

A. The kinetic energy decreases to 1/9 its original value.
B. The kinetic energy decreases to 1/3 its original value.
C. The kinetic energy increases to 9 times its original value.
D. The kinetic energy increases to 3 times its original value.

Sample Exam Answer Key

1. A	21. C	41. C	61. B
2. B	22. D	42. A	62. A
3. D	23. C	43. D	63. B
4. B	24. B	44. A	64. D
5. B	25. C	45. C	65. B
6. D	26. A	46. D	66. D
7. C	27. C	47. D	67. C
8. D	28. C	48. B	68. C
9. A	29. A	49. C	69. A
10. C	30. C	50. B	70. A
11. B	31. C	51. A	71. C
12. B	32. C	52. A	72. B
13. C	33. D	53. D	73. A
14. D	34. D	54. C	74. A
15. C	35. D	55. B	75. B
16. A	36. D	56. B	76. B
17. D	37. A	57. B	77. C
18. C	38. D	58. D	
19. B	39. C	59. B	
20. C	40. A	60. C	

Explanations for Physical Sciences Sample Exam

1. **A** is correct. The core idea of the octet rule is that the *central* atom, in this case carbon, has 8 shared electrons. This question requires critical reading of the question.

   The original atom itself usually has fewer than 8 electrons (so B is incorrect), and the passage explains that hydrogens themselves cannot have octets (so C is wrong).

2. **B** Either by familiarity with Xe and F, or by reference to the table in the text, a reader can see that 12 electrons are shared in this compound, limiting the choices to B or D. But Theory 1 only allows 8 electrons, so D is ruled out, leaving B as the correct response. This question requires the reader to examine the relationship of the passage to the conclusion given.

3. **D** The compound has 6 electrons, so it violates Theory 1 and fits Theory 2; thus A and B are ruled out. The reader is left with a choice between two structures. Since the text says that Theory 2 dictates the structure that keeps electrons farthest apart, the choice should be D. This question requires the reader to examine the relationship of the passage to the conclusion, and is particularly difficult becauses it asks the reader to think about a concept—alternate structures—only mentioned briefly in the passage.

4. **B** This answer is in fact a short restatement of Theory 2. C looks similar, but is incorrect, since it implies variability within the same atom rather than among different ones. A denies the importance of Theory 2, contradicting the sense of the question, and D contradicts the idea of greater predictive power, suggesting as it does that some experimental data are unimportant, without giving a reason. This question requires that a reader generalize from the given information.

5. **B** Theory 2 could be proven wrong by a compound having more than 12 electrons surrounding a central atom; this compound has 13. A and D are facts that fit within Theory 2, and C relates compounds fitting Theories 1 and 2 but not threatening the second theory. This question requires that a reader generalize from the information given.

6. **D** First find the molecules that have energies greater than 450 kJ/mol; then rule out those that are too long; then rule out ions. Those remaining are C_2, N_2, and O_2.

7. **C** Look at a series where the bond energy increases, e.g., Li_2 through O_2. The bond length steadily decreases.

8. **D** For example, B is true for the hydrogen ion/neutral pair, while C is true for the chlorine pair.

9. **A** According to the dot structure, N_2 has a triple bond and O_2 has a double bond; the bond energy for N_2 in the chart is also greater than for O_2. The chart shows N_2 to have a shorter bond length than O_2.

10. **C** Two electrons in a bonding orbital (one from each of the separate Li $2s$ atomic orbitals) give a single bond.

11. **B** The two $1s$ electrons on each helium atom fill both the σ $1s$ and the σ^* $1s$ orbitals, resulting in a net bond order of zero.

12. **B** N_2 has its highest electrons in bonding π orbitals. The two additional electrons necessary to form O_2 go into π^*_x and π^*_y orbitals, which are antibonding.

13. **C** Choice A predicts the correct effect, but lacks a sound reason. Choice C combines the fact that electrons in bonding orbitals tend to be between the nuclei with the observation that coulombic attraction will pull both nuclei toward the electrons in the bond, shortening the internuclear distance.

14. **D** This is true for the first, fifth, and sixth reactions.

15. **C** Combine the first reaction with each of the others (10 cells). Then combine the bottom reaction with the second, third, fourth, and fifth (4 cells). Finally, combine the second reaction with the next-to-last (1 cell).

16. **A** Here we are lowering the value of $E°$ for the Cu^{2+}/Cu cell (not the Cu^{2+}/Cu^+ reaction) by 0.337 v. We must lower all the other $E°$ values by the same amount.

17. **D** $E° = -0.13 - (0.34) = -0.47$ v

18. **C** A table of oxidation potentials would have an order that is the reverse of a table of reduction potentials. Only C reflects such a reverse order.

19. **B** Calculated potentials are:
    I. $0 - 0.34 = -0.34$ v
    II. $0 - (-0.13) = 0.13$ v
    III. no reaction; no reducing agent (e.g., Cu metal) is present

20. **C** For any value of lead ion (on the x-axis), the value of log $[CO_3^{2-}]$ is the lowest of the three anions. (Note that the y-axis shows log [anion]—not $-$log [anion], as in a pH chart —so that descending on the y-axis means going toward lower values of the anion concentration.

21. **C** The question asks for slopes. The slopes of the top and bottom lines are about equal. II would be correct if it said "about 1/2 as great . . ."

22. **D** Rule out choices A, B, and C because they do not describe inverse relationships between the two ions. Answer D correctly describes a negatively sloping line.

23. **C** For instance, take the horizontal change from -2.75 to -1.8, for which the vertical change is from -10 to -11. Then $(\Delta y)/(\Delta x) = (-1)/(.95) = -1.05$.

24. **B** The slopes for the sulfate and carbonate salts are similar, while the slope for the hydroxide is lower. One common feature of the sulfate and carbonate salts of lead is that each has one cation to one anion, where in the hydroxide there are two anions to one cation.

25. **C** This expression uses the correct rules for logarithms of products and exponents.

26. **A** Find the point where the curves for A and B cross and drop down to the line for C.

27. **C** Pressure increases. If the product were only C, the total pressure would remain unchanged (e.g. at 8 s, A is gone but C has A's initial value). But B is produced as well.

28. **C** The slopes of A and C are equal in magnitude, though not in their signs, but the slope of B is twice that of C.

29. **A** The rate of disappearance of A is a constant since the graph for A is a straight line.

30. **C** The slope of B is very high, close to 0 s.

31. **C** $P = 0.3 + 0.5 + 0.5 = 1.3$

32. **C** Note that if a concentration of a given species rises or falls linearly over time, then the reaction must be 0th-order with respect to that species; i.e., its rate of change is unaffected by the amount of the species present. But if the *log* of the concentration of the species is linear over time, then the reaction is 1st-order with respect to that concentration.

33. **D** Doubling either concentration while holding the other constant results in a doubling of the relative rate.

34. **D** Note rows 1 and 2. The change in rate from row 3 to row 4 is due only to substrate; see rows 2 and 3.

35. **D** For example, use the derived rate law with values from row 1 to get k.

36. **D** For example, use the derived rate law, where there is no dependence on the concentration of the nucleophile with row 1 to get k.

37. **A** This answer is a plausible explanation; each of the others predicts a different result.

38. **D** Note that SH^- promotes the reaction better than does OH^-, and I^- better than Cl^-.

39. **C** Kinetics studies such as these can give information about the rate-determining step, but not necessarily about other steps in the reaction.

40. **A** A body in equilibrium has constant velocity if it is in motion (and zero velocity if it is stationary). In graph A the value of the velocity is constant, that is, the slope of the curve is zero.

41. **C** For acceleration in the same direction as the motion, the velocity increases at a uniform rate. The constant slope of the velocity versus time graph gives the acceleration. The greater the acceleration the greater the slope.

42. **A** Choices C and D can be eliminated immediately because they have a slope. The slope on graph A is zero. Choice B is eliminated because it shows the velocity as constant but not necessarily zero. If the axes are distance and time, then graph A represents no change in position (no motion).

43. **D** If the velocity is low, the slope of the distance vs time curve is low because the distance covered in any unit of time is small.

44. **A** Power is the rate of doing work: $P = W/t$ where the work is force times distance:

$$P = \frac{Fd}{t} = \frac{(700 \text{ N})(5.00 \text{ m})}{3.00 \text{ s}} = 1167 \text{ W}$$

A quick estimate is given by

$$P = \frac{(7 \times 10^2 \text{ N})(5 \text{ m})}{3 \text{ s}} = \frac{35 \times 10^2 \text{ W}}{3} \sim 10 \times 10^2 \text{ W}$$

The choice closest to this value is A

45. **C** $E_p = mgh \sim (900 \text{ N})(10 \text{ m/s}^2)(3 \text{ m}) = 2.7 \times 10^4 \text{ J}$
The closest choice is C, $2.65 \times 10^4 \text{ J}$

46. **D** The person is counteracting two forces, the force of friction and the component of the weight of the box parallel to the surface of the ramp.

$$\frac{\text{parallel component}}{\text{wgt}} = \frac{\text{hgt. of plane}}{\text{length of plane}}$$

$$\text{parallel component} = \frac{3.00 \text{ m}}{5.00 \text{ m}} \times 900 \text{ N} = 540 \text{ N}$$

The total force to be overcome is 540 N + 100 N = 640 N

47. **D** The force of friction helps prevent the box from sliding down the ramp. The force required to keep the box where it is equal in magnitude but opposite in direction to the component of the mass of the box that is parallel to the surface of the ramp, minus the frictional force.

$$F = 540 \text{ N} - 100 \text{ N} = 440 \text{ N}$$

48. **B** The tube behaves like a closed-end pipe, so there is a node at the water end and an antinode at the opposite or air end of the tube. The length of the air column is one fourth the wavelength of the fundamental tone: $\frac{\lambda}{4} = X$

49. **C** Resonance occurs when the vibration of one system (tuning fork) sets up sympathetic vibrations on a near-by system (air column in tube). The second system must have a natural frequency close to that of the first vibrating system. Adjusting the tube height adjusts the wavelength, and thus the frequency of the air column, until it meets the condition for resonance.

50. **B** The wavelength of the sound depends on length of the air column producing the sound and is independent of any other medium it is transmitted into. At the boundary between the air and the water, the speed of sound changes but not its frequency. Since $V = \lambda \nu$ wavelength and speed are directly proportional. The speed of sound in a liquid is typically faster than in a gas. In order for the frequency to remain constant, at the higher speed, we expect a shift to longer wavelengths.

51. **A** By definition a complete vibration or period covers one wavelength.

52. **A** Water rises by capillary action because it "wets" the glass. The water in the tubes should, therefore, be higher than the level in the reservoir. This eliminates tubes B and C and therefore choices B, C, and D. Choice A is correct because the smaller the diameter of the tube, the higher the water will rise.

53. **D** Choice A is eliminated because the speed of all sound waves in air is constant, and independent of frequency and amplitude. Choices B and C are eliminated because the

speed is constant, the wavelength dependent on the length of the air column in the flute is also constant, and therefore the frequency cannot change. The increase in energy increases the amplitude of the wave.

54. **C** Since distance covered equals the product of the speed and the time, then:

$$t = d/v = \frac{2200 \text{ ft}}{1100 \text{ ft/s}} = 2 \text{ s}$$

55. **B** $d = v_0 t + 1/2\, at^2$ where the initial velocity $v_0 = 40$ m/s and the acceleration is due to gravity so that $a = -g = -9.8$ m/s². The negative sign occurs because the acceleration is opposite to the initial direction of motion.

$$d = \left(40\, \tfrac{m}{s}\right)(5 \text{ s}) + \left(-4.9\, \tfrac{m}{s^2}\right)(5 \text{ s})^2$$
$$= 200 \text{ m} - 122.5 \text{ m} = 77.5 \text{ m}$$

This answer is easily estimated by rounding the acceleration to 10 m/s² so that $1/2\, at^2$ becomes $(1/2)(10 \text{ m/s}^2)(5 \text{ s})^2 = 125$ m and d is approximately 200 m $-$ 125 m $=$ 75 m which is closest to choice B.

56. **B** At equilibrium, the sum of the clockwise torques must equal the sum of the counterclockwise torques. The two heavier children each sit 5 ft from the pivot point as indicated below and produce torques rotating in opposite directions.

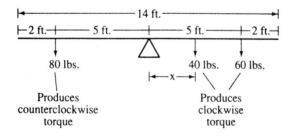

The 40-lb child must sit on the same side of the seesaw as the 60-lb child, at a distance × ft to the right of the pivot. This eliminates choices A and C immediately. Choice D is eliminated because this would form a coupled moment and equilibrium would require that there be 80 total pounds 5 ft to the right of the pivot, but the two children together have a

weight of 100 lbs. This leaves B as the only reasonable answer. Confirm this by doing the calculation:

$$(80 \text{ lb})(5 \text{ ft}) = (40 \text{ lb})(x \text{ ft}) + (60 \text{ lb})(5 \text{ ft})$$
$$400 \text{ ft-lb} = 40(x \text{ ft-lb}) + 300 \text{ ft-lb}$$
$$x \text{ ft} = \frac{(400 - 300) \text{ ft-lb}}{40 \text{ lb}} = 2.5 \text{ ft}$$

57. **B** $F_{\text{initiating}} = (\mu_s)\text{Wgt.}$ and $F_{\text{maintaining}} = (\mu_k)\text{Wgt.}$
$$F_i - F_m = (\mu_s - \mu_k)\text{Wgt} =$$
$$(0.50 - 0.20)(50 \text{ N}) = 15 \text{ N}$$

58. **D** You can save time by estimating the expected answer.
Force = Pressure × Area where Pressure $= P = 14.7$ lb/in² ~ 15 lb/in² (the area is estimated as 9 in × 11 in = 99 in² ~ 100 in²).

$$F \sim \left(15\, \tfrac{lb}{in^2}\right)(100 \text{ in}^2) = 1.5 \times 10^3 \text{ lb}$$

The best choice is D. Confirm the calculation as:

$$F = \left(14.7\, \tfrac{lb}{in^2}\right)(93.5 \text{ in}^2) = 1.37 \times 10^3 \text{ lb}$$

59. **B** Electric current, I, is the total charge per unit of time:

$$I = Q/t = 120 \text{ C}/60 \text{ s} = 2 \text{ amperes}$$

The diameter of the wire is not necessary to this calculation.

60. **C** The reaction is

$$^{131}_{53}I \rightarrow X + {}^{131}_{54}Xe$$

The other particle is emitted, not absorbed. This eliminates choice B. By conservation of matter and charge X must be $^0_1 X$ which corresponds to a positron, which is the antiparticle of an electron (zero mass number, +1 charge). Choice A is eliminated because α-particles have nonzero mass. Of choices C and D only C has a positive charge.

61. **B** For a simple pendulum the period is T and is directly proportional to the square root of the length:

$$T = 2\pi \sqrt{L/g} \text{ so that } \frac{T}{\sqrt{L}} = \frac{2\pi}{\sqrt{g}}$$

To double the period, the length must be quadrupled so that its square root is doubled:

$$\sqrt{4\,L} = 2\sqrt{L}$$

62. **A** Emission of an electron from a nucleus occurs when a neutron disintegrates into a proton and the emitted electron. Since the mass of the neutron and the proton are approximately the same, the mass number remains unchanged. The neutron is replaced by a proton so the atomic number increases by one.

63. **B** The formula is not in the simplest numeric ratio. It should read "CH_2."

64. **D** The electron in the neutral atom occupies an s orbital of higher n than the valence electrons of the positive ion. The valence electron of the neutral atom will stay, on average, farther from the nucleus than the inner electrons.

65. **B** Use $P_i = X_i P_T = (2 \times 10^{-6}) \times (1.00)$

66. **D** Mixing two bases will not result in a reaction that proceeds to a great extent since neither readily furnishes H^+. But if an acid is mixed with a base, *as long as at least one is strong*, the reaction will go essentially to completion.

67. **C** Since $K < 1$, the reaction is not spontaneous, and $\Delta G° > 0$;

$$\Delta G° = -RT\ln K$$

Although $\Delta H°$ and $\Delta S°$ combine to determine $\Delta G°$, we lack information to determine the sign of either of them.

68. **C** The oxidation state of sulfur increases from -2 to 0. Note that since H_2S is oxidized, it must be the *reducing* agent. Since the oxidation state of N in $NO_3{}^-$ drops from 7 to 4, $NO_3{}^-$ is reduced.

69. **A** Assume a 100 g sample for convenience. It will contain 5.9 g H, or 5.9 mol. It will also contain 94.1 g oxygen, or $94.1/16 = 5.9$ mol. Thus the ratio of H to O is 1:1. (The *molecular* formula—but not the empirical one—is likely to be H_2O_2.)

70. **A** Since Ag^+ is reduced at the silver electrode, that electrode is the cathode. Since electrons flow spontaneously toward that electrode, it must be positively charged.

71. **C** In fusion reactions two light mass nuclei combine to produce a heavier nucleus.

Choice B is eliminated because no nucleus was split into less massive nuclei. Choice D is eliminated because a β-particle is essentially an electron and no β-particles occur in the reaction shown. Choice A is eliminated because α-particle decay generally is not accompanied by the emission of a neutron.

72. **B** The light passing through each slit is diffracted (bent), and the two resulting beams overlap on the film to form interference patterns.

73. **A** To compare the figures, replace the prefixes by their exponential equivalents and rewrite each number in scientific notation. Therefore each becomes:
   A. $0.456 \times 10^4 \times 10^{-2}$ m $= 4.56 \times 10^1$ m
   B. 4.56×10^4 m is already in scientific notation
   C. 0.456×10^3 m $= 4.56 \times 10^2$ m
   D. $4.56 \times 10^6 \times 10^{-3}$ m $= 4.56 \times 10^3$ m

74. **A** Since the crate is moving up the incline the potential energy is increased. This eliminates choice D immediately. The potential energy cannot be determined explicitly since the vertical rise of the incline is unknown. However, the amount of work performed can be calculated.

$$W = Fd = (20 \text{ N})(10 \text{ m}) = 200 \text{ J}$$

The work was used to overcome friction and to raise the object. If the incline was frictionless, all of the work would have gone into producing potential energy. This sets an upper limit of 200 J on the increase in E_p.

75. **B** The image appears to be "inside" the mirror so the image is always virtual. This eliminates choices A and C immediately. Choice D can be eliminated because mirror images are always upright. Choice B is confirmed by tracing the rays.

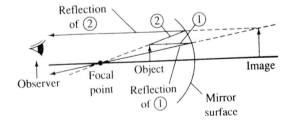

Rays parallel to the horizontal axis are reflected back through the focal point (Ray 1). Rays whose backward extrapolation would pass through the focal point are reflected parallel to the horizontal (Ray 2). The eye follows the lines of these rays back into the mirror, where they appear to converge. The result is an enlarged, upright, virtual image.

76. **B** The force of gravitational attraction is given by:

$$F_{grav} = \frac{Gm_1m_2}{r^2}$$

where G is the universal gravitational constant. The mass of the planet is $2m_1$. The radius of the planet is $2r$, so

$$F_{planet} = \frac{G2m_1m_2}{(2r)^2} = \frac{G2m_1m_2}{4r^2} = \frac{Gm_1m_2}{2r^2}$$

The gravitational force on the planet will be half that experienced on the earth.

77. **C** Kinetic energy is: $E_k = 1/2\ mv^2$. The kinetic energy is directly proportional to the square of the velocity (or speed). If the speed is multiplied by 3, the E_k must be multiplied by $3^2 = 9$.